Pocket Guide to Telecommunications, Electronic Communications, and Information Technology

Pocket Guide to Telecommunications, Electronic Communications, and Information Technology

A'isha Ajayi

Delmar Publishers

an International Thomson Publishing company I(T)P®

Albany • Bonn • Boston • Cincinnati • Detroit • London • Madrid
Melbourne • Mexico City • New York • Pacific Grove • Paris • San Francisco
Singapore • Tokyo • Toronto • Washington

Cover design by Susan Mathews, Stillwater Studio

Delmar Team
Publisher: Michael McDermott
Acquisitions Editor: Gregory L. Clayton
Developmental Editor: Michelle Ruelos Cannistraci
Production Manager: Larry Main
Art and Design Coordinator: Nicole Reamer
Marketing Manager: Kitty Kelly
Editorial Assistant: Amy E. Tucker

Printed in Canada

For more information, contact:

Delmar Publishers
3 Columbia Circle, Box 15015
Albany, New York 12212-5015

International Thomson Editores
Campos Eliseos 385, Piso 7
Col Polanco
11560 Mexico D F Mexico

International Thomson Publishing Europe
Berkshire House 168-173
High Holborn
London, WC1V7AA
England

International Thomson Publishing Gmbh
Königswinterer Strasse 418
53227 Bonn
Germany

Thomas Nelson Australia
102 Dodds Street
South Melbourne, 3205
Victoria, Australia

International Thomson Publishing Asia
221 Henderson Road
#05-10 Henderson Building
Singapore 0315

Nelson Canada
1120 Birchmount Road
Scarborough, Ontario
Canada M1K5G4

International Thomson Publishing – Japan
Hirakawacho Kyowa Building, 3F
2-2-1 Hirakawacho
Chiyoda-ku, 102 Tokyo
Japan

1 2 3 4 5 6 7 8 9 10 XXX 04 03 02 01 00 99

Library of Congress Cataloging-in-Publication Data

Ajayi, A'isha.
Pocket guide to telecommunications, electronic comunications, and information technology / A'isha Ajayi..
p. cm.
Includes index.
ISBN: 0-7668-0170-5
1. Telecommunication. 2. Computer networks. 3. Information technology. I. Title.
TK5101.A3835 1999
384—dc21 98-43746
CIP

Dedication

This book is lovingly dedicated to my best friend and colleague, Isabelle Eva Germaine Corset

Contents

Preface

The Pocket Guide to Telecommunications, Electronic Communications, and Information Technology is offered as a starting point in understanding modern technology. The goal of this text is to provide an accessible look at the emergence of these technologies, how they work, some practical applications, and future direction. Each chapter starts with an overview of related principles. Related acronyms and terms are defined to assist you in building a working vocabulary of the applications we use or have seen in our daily lives. Finally, a selected set of applications are explored, which integrate the theories and principles presented in the chapter. Easy-to-understand diagrams and illustrations are used to solidify your understanding.

ORGANIZATION OF THE TEXT

Chapter 1 – Passport to the Information Age – This chapter provides a comprehensive overview of the issues and trends that have fostered the evolution of the Information Age. A number of common scenarios are explored as a means of allowing the reader to find an entry point to the information superhighway.

Chapter 2 – The Basics – The focus of this chapter is on the fundamental concepts used in networking and other forms of electronic communications.

Chapter 3 – Getting Connected – This chapter focuses on the transmission piece of the networking equation as one of the most pivotal choices in determining the speed and subsequent quality of transmitting information.

Chapter 4 – Computers, Computers, Computers – This chapter provides the reader with a concise overview of computer basics, components, and use. Finally, a brief history of the evolution and relationship of computing to emerging and existing network models is presented.

Chapter 5 – Network Basics – This chapter is a continuation of the previous chapter. The primary focus is on the interconnection of disparate, geographically dispersed computer resources as the primary catalyst of network deployment and growth.

Chapter 6 – Telecommunications Services – Now that the reader has mastered some of the basics of telecommunications and information technology, the types of available or emerging services are presented.

Chapter 7 – The Internet – The goal of this chapter is to provide the reader with a comprehensive overview of the Internet, its evolution, and role in the emerging Information Age.

Chapter 8 – IT at Work – This chapter highlights some of the emerging applications used, or proposed for use, in various industries. The primary goal of this chapter is to demonstrate the strategic implications of these applications and services.

Chapter 9 – Managing Network Resources – This chapter discusses a comprehensive approach to managing network resources. Security is highlighted as the key element of network management.

Chapter 10 – What Every User Should Know – This chapter emphasizes the legal, regulatory, and ethical issues related to electronic communications in the 1990s.

ACKNOWLEDGMENTS

Yes, there are such things as angels:

June Joiner

Jennifer Jess

Jamie and Ray Aymerich

Connie Valois

Deb Darrow

Julie Zona

Kurt Kubitz

Ramesh Shanmuganathan

Subbiah Nagarajan

Mary Wentworth

Pygmaelion

The author and publisher thank the following reviewers who provided helpful suggestions for improving the ms:

Michael Beaver, University of Rio Grande, Rio Grande, OH

Misza Kalechman, New York City Technical College, Brooklyn, NY

Clay Laster, San Antonio College, San Antonio, TX

John Baldwin, South Central Technical College, Faribault, MN

Ron Craig, DeVry Institute of Technology, Irving, TX

Chapter 1

Passport to the Information Age

HISTORICAL OVERVIEW

Scenario 1

It's 7:00 A.M. and a gentle voice awakens you indicating that you have a meeting with the board of directors at 9:00. There are fifty-three e-mail messages awaiting your attention, five of which require immediate action. As you pedal diligently on your exercycle, a large, flat screen plays your e-mail messages complete with video, sound, and other interactive content. You reply to these messages, annotating, editing, and appending other media files as you go by speaking quietly into the air. You reorder your e-mail based on priority and conference briefly with a client in Hong Kong.

Scenario 2

As you enjoy your morning meal you watch the news and other current events, scanning the most popular electronic magazines and newspapers from around the world. Two new movies catch your eye and you schedule them for viewing when you return later that evening. An urgent reminder bursts across the screen that there is a parent-teacher videoconference at 3:00 P.M.; you have this added to schedule at the office. The electronic shopping network reviews the food and other household items that need to be replaced. As you create your electronic shopping list they are ordered from stores where these items are in stock or on sale. You make a note to pick these up from the central distribution center on your way home that evening.

Scenario 3

Wow, what a night! The varsity game resulted in a win and the ensuing celebration has left you in a daze the next morning. As you emerge to start your day, you are reminded that the term paper you have yet to begin is due in three days. You scramble earnestly and start a voice-activated search of references for your paper topic. Your printer indicates that you have 110 pages left before toner replacement is required and that the assignments due for the day are printed, collated, and stapled in the output bin. Additionally, several assignments have been queued for transmission to the appropriate instructors pending your review and selection of format digital video disk (DVD), compact disk-read-only memory (CD-ROM), or portable document format (PDF).

You go to your desk in preparation for your first distance learning class of the day. You were lucky you managed to register for a class taught by one of the most famous experts in human computer interaction in France. Oops, there is an exam today! As you answer the questions on this unexpected exercise your score is displayed in the lower right-hand corner of the screen. You breathe a sigh of relief because you squeaked by with an 88 and the instructor really liked your essay on augmented communications.

Your screen flashes a gentle reminder that you have a team meeting to complete your interactive multimedia project just after class. Darn, you forgot to create a spreadsheet of the data sent from your classmate in Australia. With a stroke or two from the keyboard, the task is done along with accompanying charts and graphics. This semester isn't going to be as bad as you thought after all.

Scenario 4

What a day! It has all gone by in a blur. You were in court this morning defending the case of an artist whose interactive performance piece is being challenged as a direct copy of a play performed on Broadway. You had to review the play on video as you got ready for work and determined that this work had, in fact, borrowed quite heavily from a television docudrama aired on cable in 1997. Upon careful investigation, you determined that most of the information and materials used in the cable version had been based on historical fact and, therefore, was in the public domain.

The attorney for the plaintiff rallied a scalding defense of his client's ownership of the material in the Broadway production, arguing that his client had taken this publicly held intellectual property and created a unique work entitling him to sole ownership. The jury had taken several hours to deliberate after being presented with electronic testimony from both sides from around the world. The judgment that followed set legal history. Since both the performance piece and Broadway play borrowed heavily from the televised docudrama, the judge ruled that both represented unique works of art and neither had a case for damages. The judge warned both parties to exercise more caution in the creative process and to be sure to respect all associated copyright and intellectual property claims used in such work.

Scenario 5

A year ago you were laid off of your corporate job in marketing. The first few months were tough as you tried desperately to find similar work. You were about to give up when it dawned on you that you owned a very powerful computer with state-of-the-art application software and high-end color peripherals. You had learned to design Web pages using a popular office suite software and decided to use your personal e-mail account to launch your own marketing research business. After the first three months, you had received more contracts than you could complete alone. So you established a network of colleagues to subcontract the various projects, creating a virtual business that billed $675,000 in the first six months!

In order to maintain quality and consistency in the completion of products, you established an electronic orientation and training system using a high-end computer as a server. Monthly status meetings and contact with subcontractors were held using the latest Motion Picture Expert Group (MPEG-4) videoconferencing techniques. You never imagined that your passion for computers and the Internet as hobbies would turn out to be the basis of your income. Go figure!

The scenarios described may seem like something out of your favorite science fiction movie or novel, when in fact most of the applications described already exist. So why do these scenarios conjure up images of the future? Unlike objects such as automobiles, forks, and other common items, the applications described in the scenarios have emerged so rapidly that we have not been able to keep pace with them and find ways to integrate them into our places of work, educational settings, or homes beyond a fax machine, networking, and Internet access. Many of us do not understand the operating principles and, therefore, the potential of these technologies remains largely unexplored.

Perhaps something much more subtle should be considered when trying to understand our apparent loath at entering the Information Age—the interaction of human with emerging technology. When we see an automobile or a fork for the first time, we are able to form a mental model of it based on metaphors that exist in the world, the experience of others, and more importantly, manipulation of the object. We are able to determine at a glance that the wheels are used to advance or move the vehicle and that some form of fuel is required to keep it going.

Yet when we sit down at a computer or try to program our video cassette recorders (VCRs) for the first time we feel awkward, stupid, or even afraid. As a result, many of us rely on others or learn enough about our computers and VCRs to perform some basic operations while some of these artifacts' enhanced features go unused or unrecognized. While a lot of this utility mentality we adopt toward technology is based on usability and design issues, a good portion of our inability to strategically deploy it is based on lack of knowledge. *Telecommunications, information technology, computer technology,* and *electronic communications* are the terms that have been associated with the applications that seem to be popping up everywhere. But what does each term mean, why are there so many abbreviations (acronyms), and why does it all seem like rocket science?

A careful analysis will reveal that the foundations of the Information Age were laid more than 500 years ago. Prior to the introduction of high-speed printing, oral information and manuscripts were the dominant mechanisms for communicating within our society. The advent of high-speed printing ushered in mass media such as books, periodicals, and newspapers. Radio, and later television, transformed existing forms of mass media to electronic communications. In the 1990s, electronic communication is a complex hybridization of wired and wireless telecommunications networks that support the high-speed transmission of voice, data, images, and multimedia.

A major paradigm shift has occurred in the ways we use and process information. Information that previously existed as tangible documents with pictures have been transformed into high-speed, digital bits and bytes and are accessed through a variety of interfaces. Many current electronic communications techniques are modeled after printing applications. At the same time, the printing industry has been transformed both in terms of its internal and operational information requirements and the varying forms printed materials have assumed.The notion of the document has evolved from data and information inscribed on paper and disseminated as periodicals, books, or other texts to the embodiment of human communications though expressed in symbols, sound, text, and images.

Recent ecological concerns and escalating paper, ink, and postage costs have been the most obvious catalysts in changing our understanding of the document. Technological developments in computing, mass storage technologies, and telecommunications have not rendered the paper document obsolete, rather it has extended the document beyond its two-dimensional limitations to an interactive, multidimensional environment where the end user may shape, control, or manipulate its contents as part of the exploration process.

EMERGING TRENDS

The advent of desktop publishing or the ability to produce high-quality documents using off-the-shelf applications software and powerful microcomputers, is perhaps the single most significant advance of the late 1980s and early 1990s. Apple Computers led the charge by developing (in conjunction with Xerox Corporation) a computer whose user interface was based on a graphical model rather than the command line interaction characteristic of the early mainframe and minicomputer environments.

Disk Operating Systems (DOS)-based IBM/Intel personal computers were more powerful, but lacked the ease of use found in Apple's microcomputers. For nearly a decade, the two platforms struggled for dominance in the various markets with IBM imbedded in many larger corporate and university environments, and Apple taking a lion's share of the Kindergarten through 12th grade markets, small-to-intermediate business users, and nearly total domination of the publishing and printing industries.

Pressure from the end user community and the need to foster global cross-platform interoperability escalated the current movement toward transparent interoperability. Many companies found themselves with a mix of IBM-compatibles and MACs, duplicate application packages, and a general inability to share data in an ever-increasingly interconnected environment.

The increase in price performance of computer peripherals such as flatbed scanners, color inkjet and laser printers, ultrahigh-resolution video monitors, and studio quality multimedia (image and video capture, sound) suites have resulted in the proliferation of computer technology into approximately 27% of all households in the United States. Software products such as Adobe's Photoshop and Premiere, Macromedia Director, and CorelDRAW place sophisticated desktop publishing tools at the fingertips of even the novice user.

Accompanying developments in mass storage devices complemented the maturing base of desktop publishing applications by providing high-capacity secondary storage for user data, files, and graphics. Magnetic storage media in the form of tape and hard drives were the earliest form of mass storage. Quarter inch cartridge (QIC) and digital audio tape (DAT) are sequential, magnetic media used primarily for backup and recovery. The sequential method of storing data on these media make them difficult to use in real-time, online applications and are best suited for archival and recovery purposes.

Large-capacity hard drives such as integrated drive electronics (IDE) and small computer system interface (SCSI) drives have made 6 megabytes (MB) the standard for most desktop configurations. Removable or portable mass storage media has undergone a number of significant changes. Bernoulli and SyQuest drives were popular within the graphic arts and printing environments, but suffered from a generalized lack of availability outside of these specialized environments. Iomega introduced several portable storage media, which capitalizes on the strengths of earlier storage technologies such as Winchester, Bernoulli, and SyQuest. Zip and JAZ drives boast portable storage capacity from 100 MB to 1 gigabyte (GB) with a price range from less than $200 to just under $400. Higher-capacity portable tape and cartridge products are available supporting storage capacities from 1–4 GB.

CD-ROM and its sister technology, the compact disk-rewriteable (CD-RW) offer all of the benefits of mass storage (600 MB+), portability, and competitive pricing. CD-ROMs are rapidly replacing floppy disks as the medium of choice for software distribution. 24X, 30X, and higher-speed CD-ROM drives provide high-speed access to photos, images, and software applications. Compact disc-recordable (CD-R) drives allow users to "burn" their own disks for in-house or external use. Digital video/versatile disk (DVD) technology combines mass storage, random access, and record capability into a single platform. In addition to these capabilities, DVD supports backward compatibility to CD-ROM, compact disk interactive (CDI) and CD-R.

Efficient electronic desktop publishing is simplified by the ability to interconnect expensive peripherals and software resources such as scanners, imagesetters, and mass storage devices with local area networks (LANs). LANs not only bring networking to the desktop, but they optimize the collaborative design and work processes by allowing users to work interactively on large, complex projects.

Connectivity for geographically dispersed resources is provided by using high-speed analog and digital transmission facilities as part of wide area networks (WANs). Since the mid-1980s, many companies have had to manage their respective telecommunications environments. During that time, networks were widely segregated on the basis of the type of data traffic they carried. It was not uncommon for organizations to have separate voice, data, and video networks complete with staff and supporting managerial resources.

Reliable narrowband communications (speeds less than 56,000 bps) soon proved inadequate. Enhancements in modulation techniques allowed more data to be transmitted across facilities designed primarily to support voice traffic by reaching speeds of 56,000 bps. While these speeds were suitable for facsimile transmission or connecting remote locations to larger networks, they could not provide the bandwidth and flexibility required for high-speed interactive or online computing.

Digital facilities provide an excellent solution for voice and data integration and later interactive multimedia data traffic in the form of voice, data, text, images, and video. It is at this point, the widespread deployment of digital transmission and technology, that we stand poised.

The cumulative impact of all the aforementioned technologies and applications stimulate end user demand. Customers were no longer satisfied with disparate networking solutions and sought a means of simultaneously transmitting voice, data, image, and video. These different data types were not to be part of separate data streams, but part of interactive documents composed of various data types. Network planners soon understood that data compression, multiplexing, and other data-encoding techniques only addressed part of the requirements for truly integrated multimedia traffic.

Integrated Services Digital Networks (ISDN) was proposed as a platform for digital integration in the United States and abroad. ISDN provided for transmission speeds in excess of 1.544/2.048 Mbps. While ISDN became wildly successful in Western Europe, it has taken nearly 15 years for ISDN to woo American users.

Network switching techniques such as Asynchronous Transfer Mode (ATM) or cell relay evolved from broadband ISDN standards. ATM provides an excellent transport mechanism for data—video, data, and voice. Another transport architecture called *frame relay*, an enhanced version of packet switching, promises data rates of 1.544 Mbps. Data is broken into variable length packets making frame relay ideal for transmitting data and images.

IMPORTANCE OF TELECOMMUNICATIONS, ELECTRONIC COMMUNICATIONS, AND INFORMATION TECHNOLOGY

Why is this area important technically and strategically to modern commerce, education, and the quality of life? Technology is a double-edged sword. On one hand, it offers many new opportunities and tools to make our work and lives easier. On the other, technology can require us to learn many seemingly impossible skills, change the content and ways in which we do our jobs, and may result in a sense of deskkilling or job elimination. How do we survive in the Information Age? While there are no easy answers, one of the ways we can learn to use emerging technology as a tool rather than a mechanism for control is to become more familiar with it.

Mastering the concepts of telecommunications, electronic communications, and information technology may seem a formidable task at first. Nevertheless, computers and other forms of information technology are becoming embedded in many aspects of our home, schools, and workplace. We are confronted with glowing light emitting diodes (LEDs), keyboards, pointing devices, and unfamiliar forms of interaction. To add to the difficulty, we must wade through a never-ending flow of alphabet soup with acronyms such as SCSI (small computer standard interface), RAM (random-access memory), and TCP/IP (transmission control protocol/Internet protocol).

The first step on this path to the information superhighway begins with knowledge. Knowledge of how information technologies work and sensitivity to the central role technology has, and will continue, to play in human development and the strategic benefits that can be derived. Let us begin with a few operational definitions of telecommunications, electronic communications, and information technology.

Telecommunications

In the 1980s, the landmark case that resulted in the divestiture or breakup of American Telephone and Telegraph (AT&T), made telecommunications a household word. Shortly after this event, devices such as fax machines, modulator-demodulators (modem), digital answering machines, caller identification (ID) and a host of wireless communications alternatives became widely available. While the term telecommunications had been used to refer to telephony (telephone communications), telegraph, and transmission facilities, the addition of computer technology in the 1950s expanded its definition to include networking and various forms of data communications.

Electronic Communications

Recognizing that the convergence of telecommunications and computers had created many opportunities and applications, new terms emerged, which emphasized the communications and messaging applications that resulted from the union. Electronic mail (e-mail), the World Wide Web (WWW), digital paging, and the portable office are just a few of the applications that we use at home, at school, and at work. Online services, enhanced videoconferencing, and networked databases added a new dimension to doing business in a global economy.

Information Technology

As computer, telecommunications, and electronic communications became more commonplace and easier to use, attention shifted from the underlying technologies to the types of applications that could be created. The term *information technology* integrates the principles of computer science and electronic communications to provide a variety of user-focused tools for home, business, and education. While many of these applications seem complex and unfamiliar, our ability to use them requires some basic understanding of how they work in order to become more productive in their use.

Chapter 2

The Basics

ANALOG AND DIGITAL BASICS —DATA AND SIGNALS

Since we have to start somewhere, the telephone seems to be our most likely candidate for exploration. Each day we place millions of telephone calls. As an artifact, we have become so accustomed to using the telephone that it takes an event such as a network outage due to natural disaster or network failure to make us aware of how dependent we have become on it. This utility mentality often leads us to take this important artifact for granted. Without the telephone, it would be nearly impossible to maintain the standard of living we enjoy and commerce as we know it would come screeching to a halt.

So how does this modern wonder of communications work? On a very basic level the telephone is an analog device. By this we mean that our voices are transmitted over a maze of wired and wireless facilities. In order to accomplish this, our voices are superimposed onto analog transmission facilities by a process called *modulation* onto analog signals. Data is placed onto the lines by varying characteristics—amplitude, frequency, and phase—of the analog line.

We will briefly introduce the concept of analog communication here and follow with a more detailed discussion in subsequent chapters. An analog wave is made up of the following components:

- Amplitude – refers to the signal height, voltage level, or intensity. Amplitude is often measured in decibels (db).
- Frequency – refers to the number of cycles or complete signal changes an analog signal assumes in a given time element. Hence, the number of hertz (Hz) per second forms the basic unit of measurement.

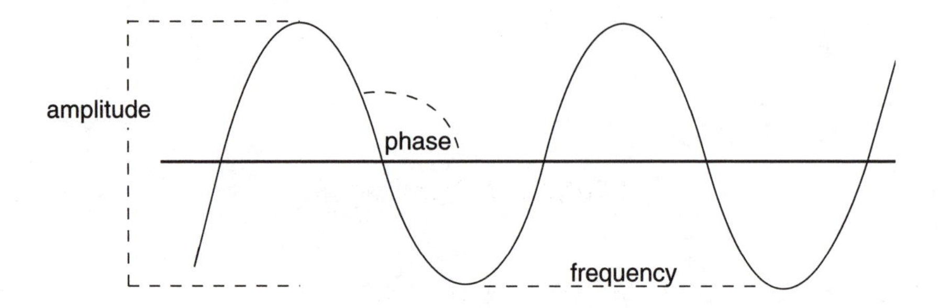

FIGURE 2-1 Analog signal

- Phase – refers to the stage or point of a signal as part of some recurring sequence; in this case, the oscillation of an analog signal over time. Phase is typically measured in degrees (°) (Figure 2-1).

By manipulating the phase, frequency, or amplitude of an analog signal we are able to transmit voice as well as data (video, text, graphics, sound, etc.) across the same lines used to carry our voice. However, the output of computers and other devices is digital. Digital devices translate data into a binary code using "0"s and "1"s. In comparison to an analog signal (where phase, frequency, or amplitude are manipulated), a digital signal uses discrete states to represent data to be transmitted.

Analog signals use amplifiers to transport data across miles of wired and wireless facilities. Afer a few miles, it is necessary to boost the signal due to attenuation (loss of signal strength as a function of distance traveled and prevailing conditions). Amplifiers are used to boost the various components of the signal, but in doing so also boost whatever noise or transient impairments that might have been picked up along the way. This is known as the "cascading effect." It is visible in cable television (TV) reception as a grainy picture or audible in a tape as hiss.

In short, after several rounds of amplification, the wanted signal is overpowered by the unwanted noise, causing an imbalance in the signal-to-noise ratio. Faxes may appear in poor quality, telephone conversations may be garbled or difficult to hear, or data is lost or corrupted as part of a modem transfer.

Digital signals are often referred to as square waves and offer far superior data integrity when compared to analog signals (Figure 2-2).

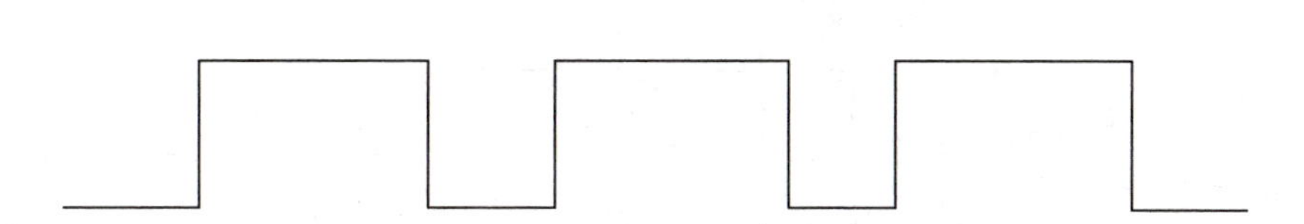

FIGURE 2-2 Digital signal

Digital signals minimize the problem of inherent and transient impairments that plague analog signals by using repeaters. Repeaters function in much the same way as amplifiers except they take a snapshot or what is called a sample of the digital signal and the data it contains. This digital representation is sent to the next repeater where it is broken down and reconstructed. Any noise that might have been picked up is discarded and the regenerated signal is transited.

MODEMS

To summarize what has been presented, data such as voice, text, video, graphics, and other types of multimedia traffic, can be carried to geographically dispersed locations over analog or digital signals. Devices such as the telephone are used to transmit our voice to almost anywhere on the globe using wired or wireless facilities. But how are other forms of data placed onto these lines? Modulator/demodulators (modems) are used to place video, text, and other multimedia data onto analog lines. Modems are a marvel because they allow us to connect to the Internet, establish dial-up connections to remote computers, and use automatic teller machines(ATMs) around the world.

In terms of composition, modems are straightforward. Most modems use a specific modulation technique or mechanism for breaking data into manageable packets, frames, or cells for transmission. These techniques will be described later in the chapter so let us focus on the device itself. Modems contain a power supply, filters, equalizers, and a unit that modulates (puts the data onto the carrier signal) and demodulates (recovers the data at the remote end). Modems may be implemented as external or standalone devices or integrated into a computer or fax machine as cards.

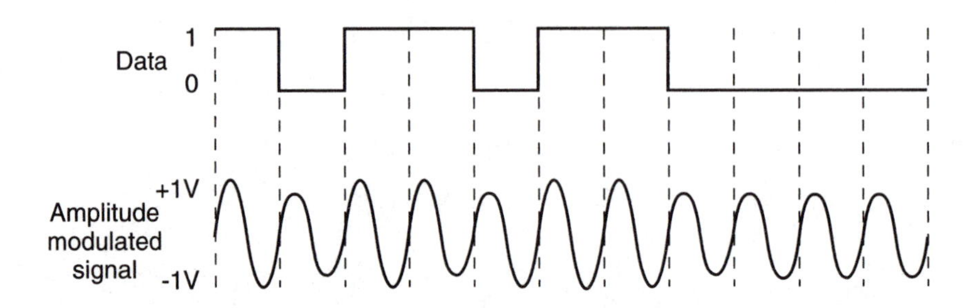

FIGURE 2-3 Amplitude modulation

Modems come in a number of varieties specialized for the applications they serve. For instance, your computer may use an external or internal modem card to connect you to the Internet. You can send and receive e-mail, download and upload files and software, or just surf the net for fun. If you have a portable office or cell phone, wireless modems could be used to send data over electromagnetic signals through the airways. Digital equivalents of modems are used to transmit data when digital lines are used.

The process of putting user data—voice, video, data—onto an analog line is accomplished by a technique called *modulation*. As mentioned earlier in the chapter, analog signals or lines have a number of components—phase, frequency and amplitude—that can be manipulated to carry user data on a line. Most present-day modems use a combination of these characteristics in order to send data over analog lines. Given the complexity of these techniques, two basic modulation techniques will be explored here.

One of the earliest modulation techniques is called *amplitude modulation* (AM) (Figure 2-3). This is the same modulation technique used to carry music over AM frequencies or radio. AM varies the height or intensity of the signal in order to represent user data. Notice that the frequency and phase remain constant whereas the amplitude varies continuously. AM is best suited for low-speed applications such as simple data transmission between ships or other vessels. As a modulation technique, it lacks the robustness necessary to handle interactive multimedia applications.

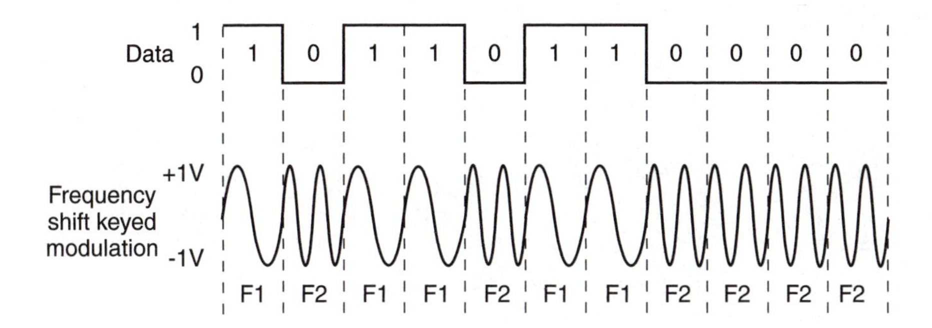

FIGURE 2-4 Frequency modulation

Frequency modulation (FM) offers a better alternative as a modulation technique (Figure 2-4). By holding the amplitude and phase constant, user data is placed onto a line by varying the frequency in accordance to changes in the user data. FM is the same technique used to carry music over FM stations. FM is excellent for high-speed data applications.

CODE SETS—ASCII, EBCDIC

In order for modems and other telecommunications devices to transmit data at high-speeds and without error, the "0"s and "1"s must be grouped into standardized units such as packets, frames, or cells. The ways in which binary information ("0"s and "1"s) are grouped and exchanged is called a *protocol.* Simply stated, protocols are based on various code sets. By grouping binary digits into packets of 7 or 8 bits in length, control information needed to keep the data, such as a résumé, intact along with the data can be represented and distinguished by the communicating devices.

This aggregating of data and control bits is not arbitrary. In fact, international standards have been adopted to facilitate global communications. One such standard or code set called *American Standard Code for Information Interchanges* (ASCII) uses a 7- or sometimes 8-bit code to represent data (Figure 2-5A).

Standard ASCII

The first 32 characters (0-31) are control codes

0	NUL	Nul
1	SOH	Start of heading
2	STX	Start of text
3	ETX	End of text
4	EOT	End of transmit
5	ENQ	Enquiry
6	ACK	Acknowledge
7	BEL	Audible bell
8	BS	Backspace
9	HT	Horizontal tab
10	LF	Line feed
11	VT	Vertical tab
12	FF	Form feed
13	CR	Carriage return
14	SO	Shift out
15	SI	Shift in
16	DLE	Data link escape
17	DC1	Device control 1
18	DC2	Device control 2
19	DC3	Device control 3
20	DC4	Device control 4
21	NAK	Neg. acknowledge
22	SYN	Synchronous idle
23	ETB	End trans. block
24	CAN	Cancel
25	EM	End of medium
26	SUB	Substitution
27	ESC	Escape
28	FS	Figures shift
29	GS	Group separator
30	RS	Record separator
31	US	Unit separator
32	SP	Blank space (Space bar)

33 !
34 “
35 #
36 $
37 %
38 &
39 ‘(
41)
42 *
43 +
44 ,
45 -
46 .
47 /
48 0
49 1
50 2
51 3
52 4
53 5
54 6
55 7
56 8
57 9
58 :
59 ;
60 <
61 =
62 >
63 ?
64 @
65 A
66 B
67 C
68 D
69 E
70 F
71 G
72 H
73 I
74 J
75 K
76 L
77 M
78 N
79 O
80 P
81 Q

82 R
83 S
84 T
85 U
86 V
87 W
88 X
89 Y
90 Z
91 [
92 \
93]
94 ^
95 _
96 `
97 a
98 b
99 c
100 d
101 e
102 f
103 g
104 h
105 i
106 j
107 k
108 l
109 m
110 n
111 o
112 p
113 q
114 r
115 s
116 t
117 u
118 v
119 w
120 x
121 y
122 z
123 {
124 |
125 }
126 ~
127

FIGURE 2-5A ASCII

Character	EBCDIC	ASCII	Character	EBCDIC	ASCII
A	1100 0001	100 0001	S	1110 0010	101 0011
B	1100 0010	100 0010	T	1110 0011	101 0100
C	1100 0011	100 0011	U	1110 0100	101 0101
D	1100 0100	100 0100	V	1110 0101	101 0110
E	1100 0101	100 0101	W	1110 0110	101 0111
F	1100 0110	100 0110	X	1110 0111	101 1000
G	1100 0111	100 0111	Y	1110 1000	101 1001
H	1100 1000	100 1000	Z	1110 1001	101 1010
I	1100 1001	100 1001	0	1111 0000	011 0000
J	1101 0001	100 1010	1	1111 0001	011 0001
K	1101 0010	100 1011	2	1111 0010	011 0010
L	1101 0011	100 1100	3	1111 0011	011 0011
M	1101 0100	100 1101	4	1111 0100	011 0100
N	1101 0101	100 1110	5	1111 0101	011 0101
O	1101 0110	100 1111	6	1111 0110	011 0110
P	1101 0111	101 0000	7	1111 0111	011 0111
Q	1101 1000	101 0001	8	1111 1000	011 1000
R	1101 1001	101 0010	9	1111 1001	011 1001

FIGURE 2-5B EBCDIC and ASCII

ASCII is currently the most commonly used code set for microcomputer communications. There are other more powerful code sets in existence such as Extended Binary Coded Decimal Interchange Code (EBCDIC) was developed by IBM for use in its mainframe computing environment. Many mainframes and other computer environments use this code set. In order for computers with different code sets to interact in a networked environment, protocol conversion is used to make the necessary translations (Figure 2-5B).

When we place a voice telephone call, we do not have to worry about timing. One of the connected parties speaks at a time. However, when we send text, files, video, or graphics things get a bit more complicated. When we send a copy of a résumé we must maintain the integrity

Start bit	0110101	Parity bit	Stop bit

0110101 = One character

FIGURE 2-6 Asynchronous communication

of the document. Characteristics of the document such as page size, orientation, the beginning and end of the document, margins, spacing, font, and other elements of the résumé are translated based on the code set and protocol used. Timing is used to maintain flow on the transmission path and keep the packets from being lost, out of sequence, or intermixed with packets from other documents on the network.

ASYNCHRONOUS AND SYNCHRONOUS COMMUNICATIONS

There are two mechanisms used to establish timing or *clocking,* as it is more technically referred to, of data as part of the flow of traffic on a network—asynchronous and synchronous. In asynchronous communications, start and stop bits are added to the data stream to help distinguish the beginning and end of the transmission and the packets that are part of a given transmission and to maintain flow (Figure 2-6). While asynchronous communication, in conjunction with ASCII as the code set, is the most widely deployed combination in modem communications it is relatively ineffective for very high-volume, speed data transfer. The start and stop bits along with the other control bits used by the ASCII increase as a function of the number of data packets sent. Essentially, as the volume of the data increases, the amount of control information or overhead needed to transmit it increases. The result is most of the bandwidth available or information-carrying capacity of the line is consumed by control information, reducing the amount available for the user's data.

Synchronous communication solves this by grouping multiple characters into blocks (Figure 2-7). Asynchronous communications transmit one character at a time. By grouping characters into blocks, less control information is required to send more data across the line. Instead of using

Flag	Address	Variable length data	FCS	Flag

FIGURE 2-7 Synchronous communication

start and stop bits to establish and maintain timing, clocks in modems and other networking devices are used. Synchronization bits are sent to keep these clocks in synch, thus maintaining the flow of traffic on the line. While synchronous communication is a more efficient approach, associated devices are more expensive due to increased circuitry and intelligence.

BITS, BYTES, PACKETS, AND FRAMES

In addition to the information covered, it is important to note that the speed and efficiency of a given transmission technique or protocol has a lot to do with the ways in which the data and control information are packaged. The bit or binary digit is the most fundamental unit for transmission. Bits are aggregated into bytes or a group of 8 bits for transmission. Depending on the character set, a byte may be used to represent a single character. Characters are grouped into blocks or words. Blocks are placed into fixed or variable-sized frames and so on for transmission. There is a delicate balance that must be maintained between packet size, protocol, and the quality of the subsequent transmission. Consequently, larger packet or frame sizes do not necessarily equate to more data sent.

SIMPLEX, HALF DUPLEX, AND FULL DUPLEX

Now that you have mastered the basics of how information is formatted for transmission, we will discuss direction. Not only must we be concerned with the rate of flow of a given transmission, it is important to maximize the speed and efficiency of a given transmission path by matching it to the requirements of a given application. For example, when utility companies remotely read meters for gas and electricity, the

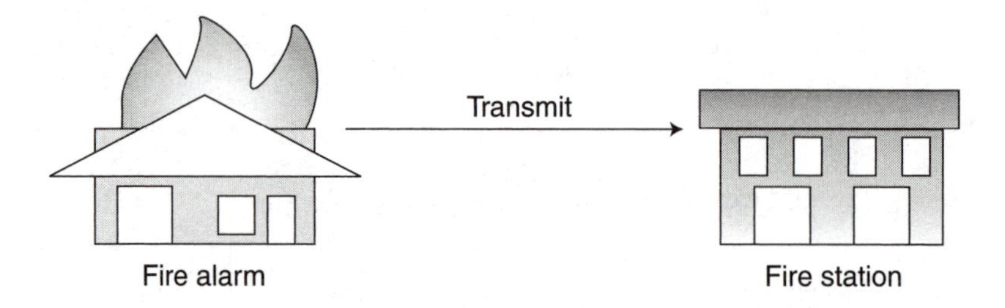

FIGURE 2-8 Simplex transmission

information only needs to travel in one direction. This unidirectional flow of data across a link is called *simplex* (Figure 2-8). Simplex transmission is best suited for applications that do not require feedback.

When an alarm goes off and it is attached to a monitoring center or service, information about the nature and location of the breach must be verified. In these instances, a simplex communication is inadequate. A path is necessary to receive information from the point of the alarm and a path is required to transmit system queries and receive status informa-

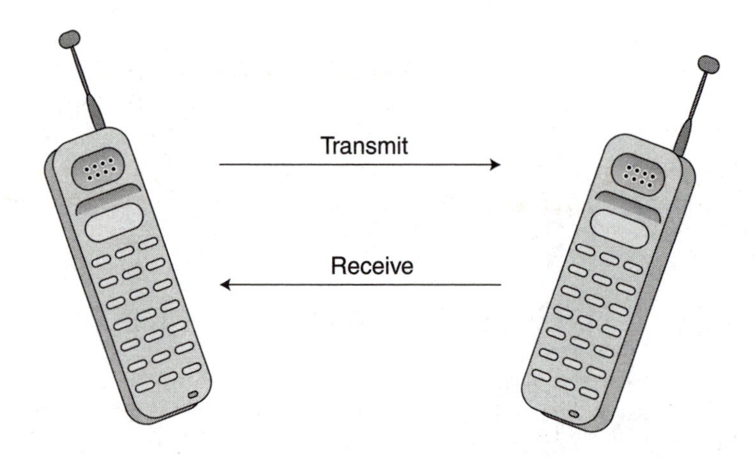

FIGURE 2-9 Half duplex (HDX) transmission

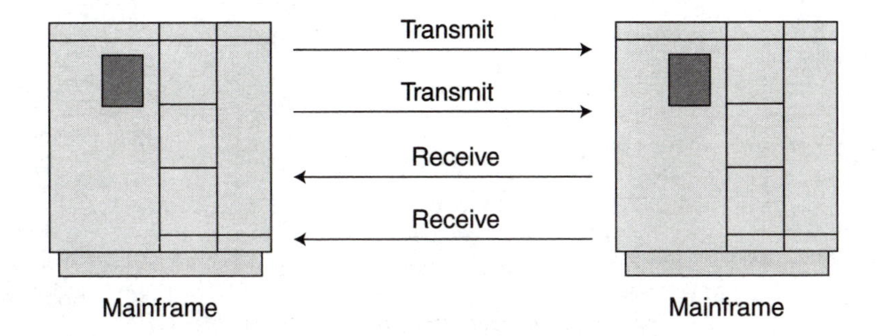

FIGURE 2-10 Full duplex (FDX) transmission

tion. The use of dual paths over a single transmission facility is called *half duplex* (Figure 2-9). In half duplex or HDX transmission, two wires are typically used: one to transmit and one to receive. As data is transmitted over one path, the other path remains idle waiting for the signal to change direction or perform what is called *line turnaround*. HDX transmission is used for the application described as well as the first portion of voice or data transmission.

You might be asking yourself, why does the other path remain idle? Also, does not line turnaround have an impact on transmission time? Yes, HDX can be relatively inefficient when there are large amounts of data to be transmitted from numerous communicating entities. By creating four paths within a given transmission facility, two can be used for transmitting and two to receive data simultaneously. As a result, more data can be supported at much higher rates. Full duplex or FDX transmission is used to carry multiplexed (or combined) signals from multiple HDX telephone calls onto larger FDX transmission paths (Figure 2-10).

MULTIPLEXING

Transmission facilities are perhaps the most expensive portion of network costs. In order to maximize transmission facilities, a technique called *multiplexing* is used to put multiple user data on the same line. Multiplexing is based on an awareness of timing and frequency. All transmission facilities, whether they are wire such as telephone lines or wire-

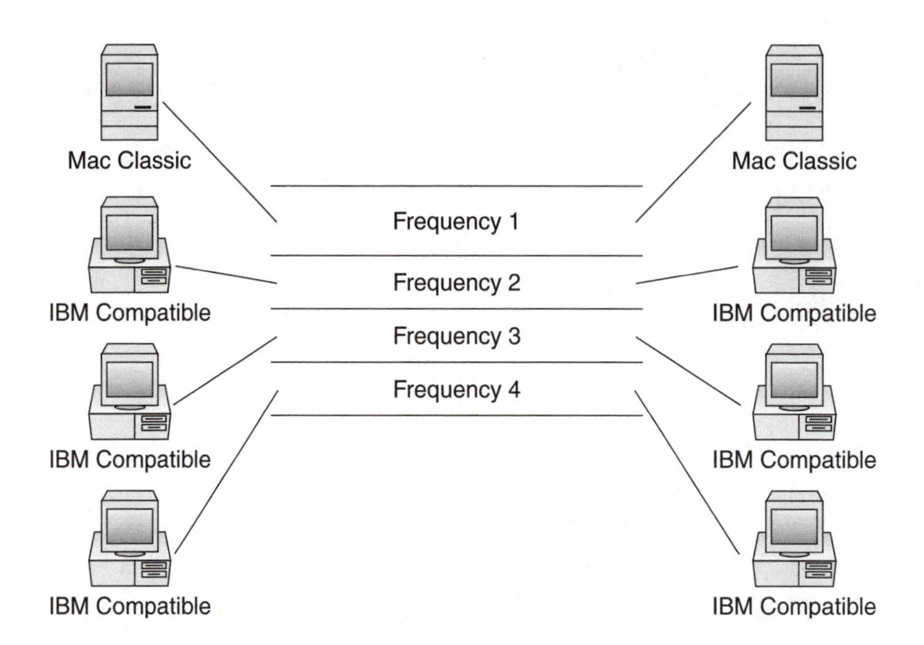

FIGURE 2-11 Frequency division mulitplexing (FDM)

less such as radio frequencies used to carry music, video, or cellular traffic, have a range of frequencies available to carry data. This capacity or range of frequencies is referred to as the bandwidth of a given medium.

Available bandwidth can be divided into separate frequencies separated by guardbands. In this instance, the data from various user devices is placed on a distinct range of frequencies. This technique is called frequency division multiplexing (FDM) (Figure 2-11). FDM is used to share facilities for multiple voice telephone calls in the public switched telephone network (PSTN). FDM is also used to carry multiple radio signals over frequencies allocated by the Federal Communications Commission (FCC) for FM broadcast. Broadcast TV and cable TV deploy this technique as well.

Additionally, each line can be exploited by staggering the timing of user data as it travels across the link. By assigning the packets or frames from each user device a time slot, a single line can carry multiple user traffic. The use of time slots to carry multiple signals is called time divi-

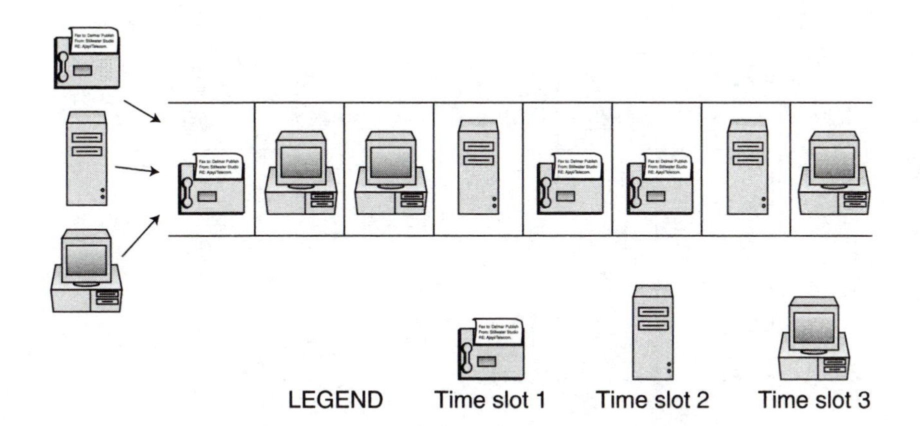

FIGURE 2-12 Time division multiplexing (TDM)

sion multiplexing (TDM) (Figure 2-12). TDM represents a major enhancement over FDM where a lot of bandwidth is wasted on guardbands.

TDM is not without its flaws. Once the time slots are assigned they are fixed. In short if a device has no data to send, the time slice allocated is still reserved and goes unused. Synchronous TDM overcomes the problem of fixed time slots by using dynamic resource allocation. In many instances, both FDM and TDM are used together. First, the available bandwidth of a given facility is divided into discrete frequencies. Next, each frequency is divided into time slots, taking advantage of more of the line's information-carrying capacity (Figure 2-13).

PUTTING IT ALL TOGETHER

Now that you have completed this chapter, what are the important concepts to remember? First, data from either voice or data devices can be carried using either analog or digital transmission. Analog works well for low-speed communications or narrowband—speeds below 56,000 bits per second. Analog transmission is used extensively for personal computing and associated applications, whereas modern enterprises require the speed and efficiency offered by digital transmission. The decline in prices and the generalized availability of digital transmission alternatives will likely result in ubiquitous digital transmission in the near future.

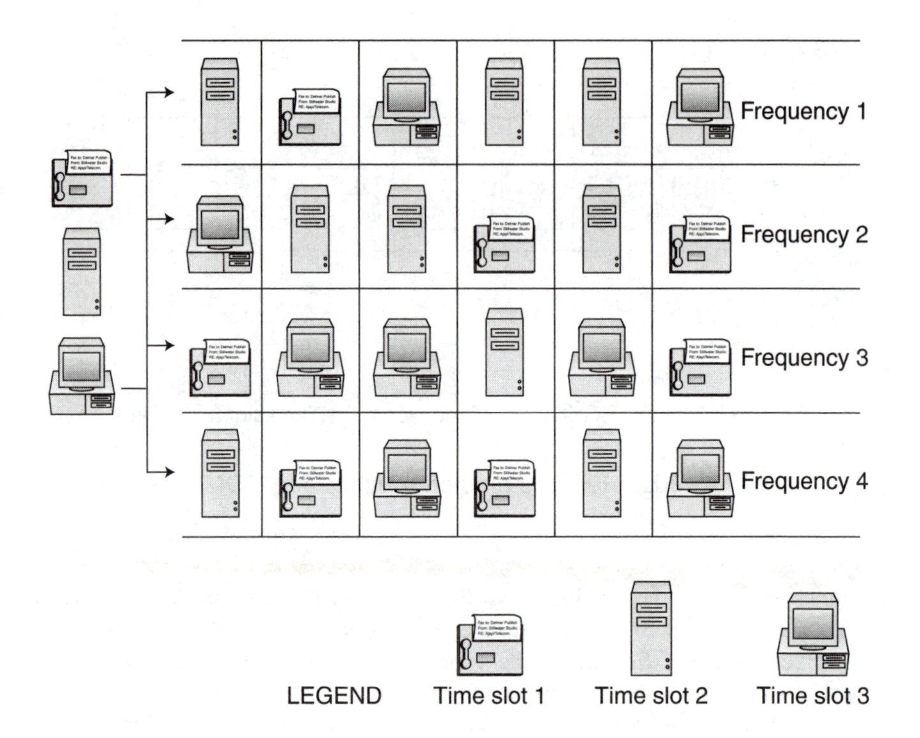

FIGURE 2-13 TDM/FDM

In order to maximize transmission efficiency on a given link or line, data must be aggregated into standard units of bits, bytes, blocks, and frames for transmission. The form that data assumes as it is transmitted is standardized by using a specific code set or protocol. ASCII and EBCDIC were presented and compared as two popular code sets for data transmission. ASCII still dominates personal computing whereas EBCDIC and other more powerful code sets and protocols are used to support business communications.

While the telephone is used to transmit and receive voice communications, devices such as modems are used to carry data, video, and other forms of interactive multimedia traffic. Modems are used to superimpose data onto, and recover from, the medium of choice. Modems may be external or internal and vary in terms of speed and sophistication of flow control, error recovery, and other factors that contribute to transmission efficiency.

Timing or synchronization was reviewed as another contributing factor to transmission efficiency. In one instance, ASCII is used with asynchronous communications for most personal computer (PC)-based communications. While asynchronous communication is sufficient for PC communications, the volume of traffic required by most business applications involve communication with larger computers such as mainframes and minicomputers. In those instances, synchronous communications provide better use of available transmission facilities.

The flow of data transmission is another important factor in maintaining high data rates. A single path is used for simple monitoring applications while more complex functions such as alarm systems and cable television support more complex functions through the use of HDX and FDX facilities.

Finally, multiplexing is used to maximize transmission efficiency. FDM and TDM are two techniques used extensively in modern data transmission. By combining all of the elements outlined in this chapter, we have the basis of modern high-speed networks capable of transporting interactive multimedia traffic.

Chapter 3

Getting Connected

OVERVIEW OF COMMUNICATIONS CIRCUITS

The primary goal of information technology is to provide the user with a fast and reliable means of transmitting information between geographically dispersed locations. Many factors must be considered in trying to decide which facilities are most suitable for a given transmission requirement. In addition to speed, distance, the choice between analog or digital facilities, and network management, we must also understand the nature of the amount and flow of traffic.

Transmission involves the movement of data and information between geographically dispersed locations. There are several key goals of efficient transmission

- Speed – A balance between multi-megabit data transmission and cost.
- Reliability – Consistent results in transmitting error-free high-speed data.
- Availability – Access to facilities when needed due to minimum downtime or traffic overload.
- Flexibility – A range of alternatives in terms of data traffic supported (voice, text, graphics, video, or multimedia) and facilities (analog or digital, switched or leased).
- Security – Data integrity and system protection.

An intricate web of transmission facilities provides connectivity to needed resources. These facilities are alternately referred to as lines, circuits, or paths. This web or network of transmission facilities is made up of the following elements:

1. Wired and wireless – Wired facilities are physical links between attached nodes and devices. Copper twisted pair, coaxial cable, and fiber optics are the most commonly used wired connections in LANs, WANs, and Global Area Networks (GANs). Wireless facilities such as microwave, cellular, and satellite provide connectivity by using various segments of the electromagnetic spectrum. Hence, physical wires are eliminated. Whereas wired and wireless communication facilities have evolved through parallel phases, wired facilities offer the user greater flexibility and control over the characteristics of subsequent transmission. Wireless communication has proven its worth in providing connectivity in areas where physical or environmental issues make it impractical, if not impossible, to install wired facilities.

2. Leased and switched – Leased lines provide permanent or semi-permanent connectivity between high-volume locations. With the use of switched facilities network nodes and resources are connected on an "as needed basis." Essentially, the call path and required resources are configured and used for the duration of the transmission much in the same way typical telephone calls are made.

3. Analog and digital – Analog transport data on a continuously varying signal. Analog transmission dominated voice communication and remains a viable option for connectivity in LANs, WANs, and GANs carrying data traffic (text, graphics, video, and multimedia). Digital facilities are becoming the backbone of modern communications due to its speed, security, and data integrity. Digital facilities are often used to form the long haul trunks or paths between nodes and other resources.

4. Public and private facilities – Public transmission facilities are provided by a carrier or company with legal authority to provide coverage and transmission service within a given area. These providers or common carriers offer an array of connectivity alternatives for users. Private facilities or in-house facilities comprise networks and associated resources that are "owned" by a user or enterprise.

By combining the aforementioned elements, users have access to a staggering array of connectivity alternatives. This section will address each of these characteristics in more detail and finally compare them in terms of the benefits and considerations that must be addressed prior to selecting the optimum mix for a given organization.

Using Leased and Switched Lines

Most networks are made up of both dedicated and switched facilities. For example, the public switched telephone network, which is made up of local and long distance services, uses switched facilities from customer or subscriber locations to central office facilities and then leased lines are used between central office switches across the United States.

Most CATV or cable TV networks use dedicated facilities to connect our homes to the network. In this example, users are always connected to the head end of the cable company even if their sets are not turned on. There is also an economic justification for deciding when to use switched or leased facilities. Switched facilities are excellent for providing ad hoc access to advance telecommunications applications and resources. For example, companies that have an occasional requirement for videoconferencing do not have to invest in onsite videoconferencing facilities and associated staff. They can simply call a common carrier or other service provider and use videoconferencing on an "as needed basis."

LINE CONFIGURATIONS

Now that you understand the ways in which lines can be used, we then turn our attention to the ways in which geographically dispersed locations are interconnected. There are a number of interconnection options or line configurations currently in use. They include point-to-point, multipoint or multidrop, or a mixture of these facilities and what we call a hybrid or meshed network. Each line configuration varies in terms of speed, cabling costs, reliability, security, and other important network management issues.

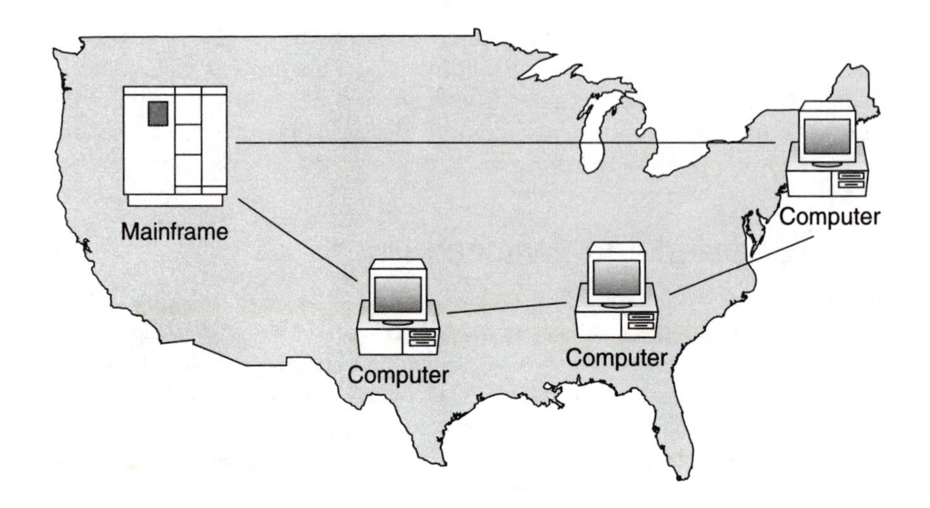

FIGURE 3-1 Point-to-point configuration

Point-to-Point

Most modern networks are made up of point-to-point facilities that have grown as the company transmission needs have changed. Point-to-point facilities are simple. For example, if we wish to transfer data between two locations, all that would be required to connect the two locations is a line between them (remember, this line may be switched or leased) and a modem in each site (Figure 3-1).

Multidrop or Multipoint

As the number of connections grows, a simple point-to-point configuration can be difficult to manage and be extremely costly. By attaching multiple locations to common facilities or in a multidrop configuration, each location may have its own modem or may share modems in what is known as a modem pool. Multipoint or multidrop configurations are the best way to provide the best connectivity between multiple sites while minimizing equipment and transmission costs (Figure 3-2).

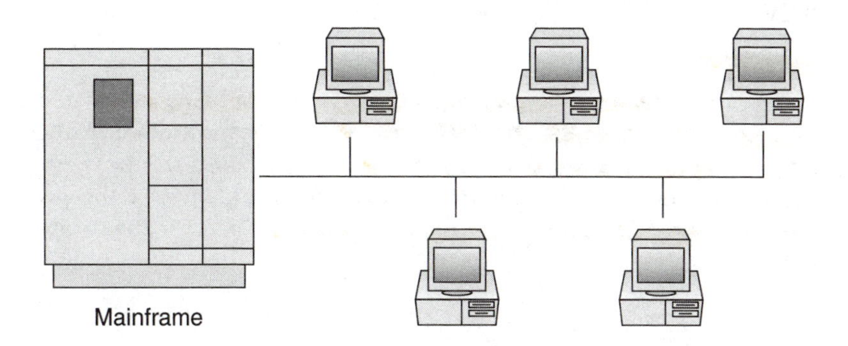

FIGURE 3-2 Multipoint or multidrop configuration

Combining Point-to-Point and Multipoint

Given the need to support voice, data, image, and other forms of interactive multimedia traffic, most private and public networks are made up of a combination of switched and leased facilities that are implemented using both point-to-point and multidrop configurations. By combining these elements, these networks provide flexible, high-speed infrastructure to carry a variety of data traffic to meet an ever-increasing array of networking needs.

ANALOG AND DIGITAL FACILITIES

So far we know that switched or dedicated lines can be interconnected in either point-to-point or multipoint configurations. From the previous chapter we know these facilities can also be implemented as analog or digital.

For more than 100 years, analog communications dominated the information landscape. In its simplest form, analog communication involves the transport of data over a carrier or signal that is modulated. The process of modulation may involve conversion of digital input data onto an analog signal. An analog signal is characterized as an oscillating or continuously varying signal.

Key Concepts Review

Carrier

A carrier is a continuously varying signal capable of being modulated or impressed with a second data-bearing signal. The bandwidth and, therefore, the available range of frequencies will vary based on the medium deployed. The carrier is typically the center frequency of the given medium; signal strength is strongest toward the center of the available frequencies.

Modulation

Modulation is the process by which some characteristic of one signal is changed according to some characteristic of an input signal. In analog communications the amplitude, frequency, or phase of a signal may be manipulated singularly or in combination to transport data.

Demodulation

Demodulation is the process by which data and control information is recovered and routed to its ultimate destination(s). It may involve reconverting a signal from analog to digital and the execution of an error detection routine prior to sending data to the destination device.

Bandwidth

Bandwidth refers to the information-carrying capacity of a given transmission facility. Bandwidth is a metric of how much data can be transported during some specified time period—typically one second.

Frequency

Frequency is the rate of oscillation, the number of cycles per second, measured in hertz.

Modulation Techniques

There are many modulation schemes used to superimpose data onto a path or link. Most of these methods are based on, or derived from, amplitude modulation, frequency modulation, or other more sophisticated modulation techniques.

> *Amplitude modulation* (AM) – The process of superimposing or modulating data onto a link by varying the amplitude of the carrier signal.

> *Frequency modulation* (FM) – The process of superimposing or modulating data onto a link by varying the frequency of the carrier signal.

Multiplexing Techniques

In 1874 Alexander Graham Bell's invention of the harmonic telegraph and other related developments provided a means to transport multiple signals simultaneously over the same transmission facility or what is also known as multiplexing. Multiplexing is both an economic and a technical necessity in modern networks. The need to reduce costs associated with transmission facilities and the need to reduce physical cabling or plant, are accomplished by using a single transmission path to carry multiple signals. This is accomplished by assigning input signals to different frequencies, time slots, or both on the same facility.

Frequency Division Multiplexing

The available range of frequencies of a given transmission facility are divided into smaller chunks or subchannels. Each input signal is placed on one channel with a gap or range of non-data-bearing frequencies called a *guardband* to separate each multiplexed frequency.

Time Division Multiplexing

Using TDM, time slots are allocated to different channels for each of the input signals. Fixed cycles or frames are used to provide access to the total available bandwidth for a given time interval.

A hierarchy of TDM channels and associated bandwidth have been adopted by North American and European standards-setting organizations. This hierarchy is typically applied to digital transmission facilities such as T1 or ISDN. This will be covered in detail in Chapter 6.

Digital principles

Originally, analog transmission facilities were optimized to support narrowband voice communications or transmission rates <56,000 bps. By using intelligent network devices such as modems, concentrators, routers, and bridges, analog facilities were able to transport image, data, and slow scan video over geographically dispersed network resources. However, inherent and transient impairments, data security, and low data rates soon made digital transmission a more attractive alternative. Digital transmission facilities were developed by AT&T in the late 1960s for use in its long distance network. Within a decade competitive digital offerings forced AT&T to incorporate digital transmission facilities capable of supporting broadband communications or data speeds in excess of 56,000 bps into its customer offerings.

Since the output of most devices in a network is digital, logically the best transmission scenario—optimum speed, data integrity, traffic integration, and security—would be served by end-to-end digital communications. However, many networks carry analog data translated by modems from digital input signals. A device called a *codec* is used to convert these analog signals to digital where they are transmitted over the network. By converting these analog signals to digital, it is possible to integrate voice, data, video, and other forms of interactive multimedia traffic over common digital facilities.

Data compression techniques provide a means of reducing the amount of bandwidth needed to transmit data. Adaptive Differential Pulse Code Modulation (ADPCM) dynamically converts the differences between samples rather than absolute values as in PCM. ADPCM allows the transmission of voice grade signals using 16 Kbps–32 Kbps resulting in increased overall throughput over digital transmission facilities.

Joint Photographic Experts Group (JPEG) compression standards have been used to divide an image into pixel blocks that can be halved. The total number of times this halving can be tolerated will depend on the desired amount of image quality required for the image once it is decompressed. Full-motion video can be transmitted over low-speed analog and high-speed digital facilities by using the Motion Picture Experts Group (MPEG) compression standard. MPEG-1 provides CD-quality audio and a standard image of 15-bit color, 352 x 240 with a frame rate of 30 frames per second (fps). The MPEG-2 standard is used for broadcast-quality video over broadband facilities.

Analog and digital facilities are available for data transmission. With the arrival of digital communications many predicted that our dependence on cheaper, slower analog facilities would diminish. The dominance of analog facilities for over a century has resulted in a situation where many enterprises have amassed large investments in supporting analog transmission facilities and technology. While digital transmission offers users more flexibility in terms of speed and transmission quality, they are typically more expensive and costlier to maintain than analog networks.

WIRED AND WIRELESS COMMUNICATIONS FACILITIES

We have seen that there are a number of factors to consider in building network infrastructure. Of all the costs associated with modern networks, it can be said without qualification that the underlying infrastructure made up of the circuits or paths, whether they are point-to-point, multipoint, analog or digital, or switched, is the most important and potentially most costly. Few enterprises have the luxury of constructing a network from scratch. Typically, there is an embedded mature base of facilities that make up these networks. While most of the attention has focused on wired networks, network connectivity using wireless facilities or radio frequency have been used for almost the same time period.

There are a number of ways to classify modern transmission facilities. We can classify them in terms of physical circuits made up of copper or fiber. We may also look at circuits used to provide connectivity to modern networks in terms of its use of the electromagnetic spectrum or radio waves. In the first case, when a physical medium is used to connect multiple locations we refer to this as guided facilities. When geographically dispersed facilities are connected using radio frequency (RF) we referr to this as *wireless* or *unguided*. Therefore, communication circuits or facilities can be grouped in terms of whether they are guided or unguided.

Guided

Guided or wired facilities make up the majority of infrastructure in the public switched telephone network and most private networks. Over the past 100 years, guided transmission facilities have involved copper to

FIGURE 3-3 Copper twisted pair cable

ultrapure glass or fiber optics. The three media that are commonly used to implement wired facilities are copper twisted pair, coaxial cable, and fiber-optic cable.

Copper Twisted Pair

Copper twisted pair is perhaps the oldest of the three guided or wired transmission facilities in use (Figure 3-3). At one point, copper twisted pair made up the bulk of the facilities used to implement the PSTN. Remember that the PSTN is made up of both local (telco-based) and long distance (LDS) service. Copper is an excellent medium for transmitting electrical signals. Through the process of modulation we are able to carry data on these signals over copper facilities. Two strands of copper wire are twisted to minimize certain characteristics or impairments that plague copper twisted pair. These impairments include electromagnetic interference (EMI), cross talk (for example, other telephone calls may be heard on the same line), and attenuation.

Perhaps the chief benefits of copper twisted pair lies in its cost, ease of implementation, and flexibility. Copper twisted pair is a popular medium in local area networks or LANs such as EtherNet or other local networks that use TCP/IP. The chief limitations of copper twisted pair lie in the amount of traffic that can be carried at very high speeds, due to potential interference.

Coaxial Cable

Coaxial cable solves some of the problems characteristic of copper twisted pair by using a larger copper core surrounded by shielding (Figure 3-4). While coaxial cable is less susceptible to the impairments that plague copper twisted pair, its higher cost is offset by increased

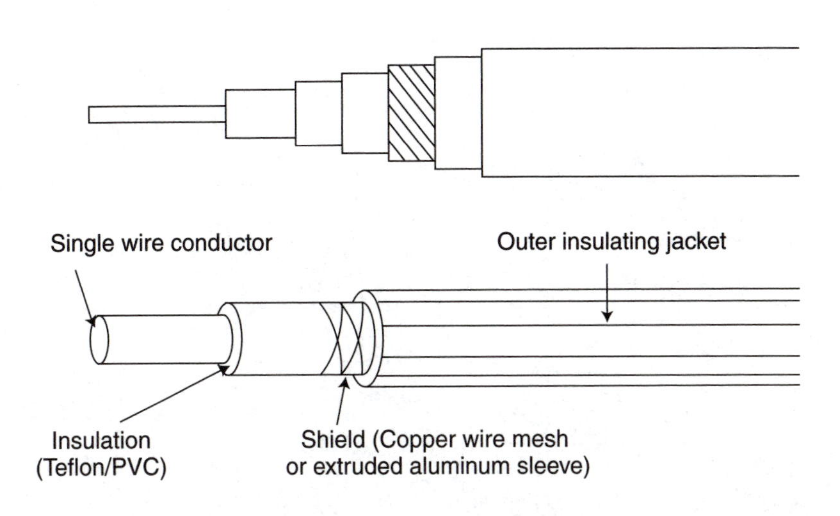

FIGURE 3-4 Coaxial cable

speed, reliability, and bandwidth. Coaxial cable is used extensively in CATV networks, LANs, and in the past, some remote portions of the PSTN.

Fiber-optic Cable

Both copper twisted pair and coaxial cable are excellent media for data transmission. However, their respective capabilities in meeting the demands of modern networking—high-speed, error-free transmission—have reached an upper limit. By using laser (light amplification by stimulated emission of radiation), LED (light-emitting diodes), and infrared as sources, fiber-optic transmission boasts virtually error-free gigabit transmission rates. Fiber-optic cable can be made of glass or plastic. Unlike copper twisted pair or coaxial cable, fiber-optic transmission does not suffer from electro-magnetic interference (EMI). In short, pulses of light are used instead of electrical signals to convey data. Fiber optics is best deployed in networks where multiple types of traffic must be carried over great geographic distances. Fiber-optic cable is used extensively in the long distance portion of the PSTN, high-speed LANs such as fiber distributed data interface (FDDI), and in CATV networks (Figure 3-5).

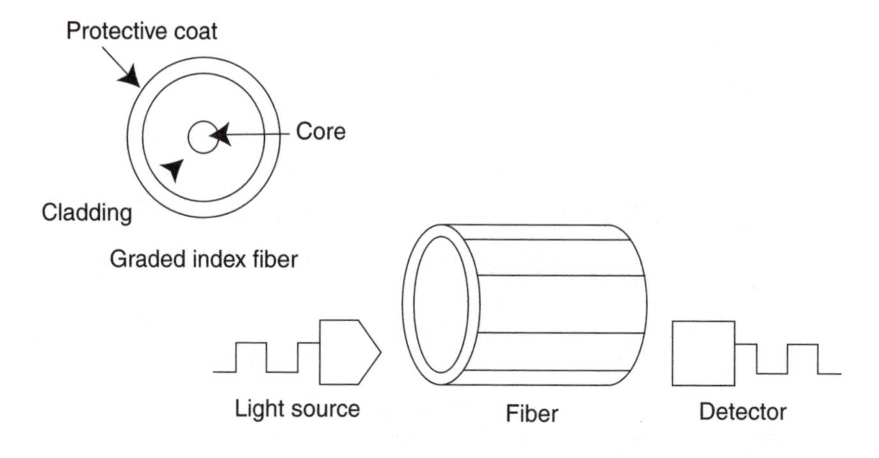

FIGURE 3-5 Fiber optic cable

Unguided or Wireless Facilities

Wireless communications implemented in the form of various radio frequency-based technologies have long served as an adjunct to wired networks. Wireless communications offer flexibility for implementation over areas where it would be impossible to lay cable or in highly developed metropolitan areas. Though there are many different types of wireless communications in existence, we will examine microwave, cellular, satellite, and personal communications systems.

Microwave Communications

Microwave communications provide data transmission rates in excess of 1.54 Mbps. Using terrestrial microwave dishes, data can be transmitted between locations where the distance between them is not >25 miles. This 25-mile limitation is known as line of sight (LOS) (Figure 3–6). Microwave communications can support both analog and digital transmission. Microwave is used to implement T1 and other broadband digital transmission offerings. In earlier years, microwave was used to implement long distance service. Faster, more reliable media such as fiber-optic cable has caused microwave to be used primarily as a backup facility.

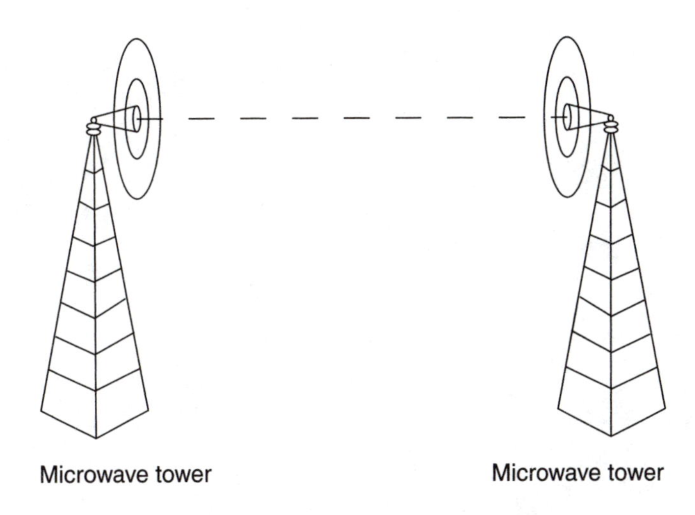

FIGURE 3-6 Line of sight in a typical microwave system

Microwave communications provide an excellent means for implementing high-speed analog and digital networks. It is more costly than copper-based transmission and is susceptible to many of the same impairments. Attenuation, absorption, and multipath fading are a few of the impairments that impact this form of transmission.

ATTENUATION. This is the tendency of signal strength to diminish as a function of distance. As signal travels across a link, its power or amplitude weakens. Wireless communications are especially susceptible to attenuation.

ABSORPTION. Adverse weather conditions such as fog, rain, and snow can interfere with various forms of wireless transmission. The strength of the signal is distorted as it is absorbed by atmospheric conditions.

MULTIPATH FADING. As the signal leaves the microwave dish, it fans out causing a portion of the signal to be reflected back into the originating source. These wireless signals are reflected from the ionosphere or flat surfaces on the earth causing signal degradation and errors.

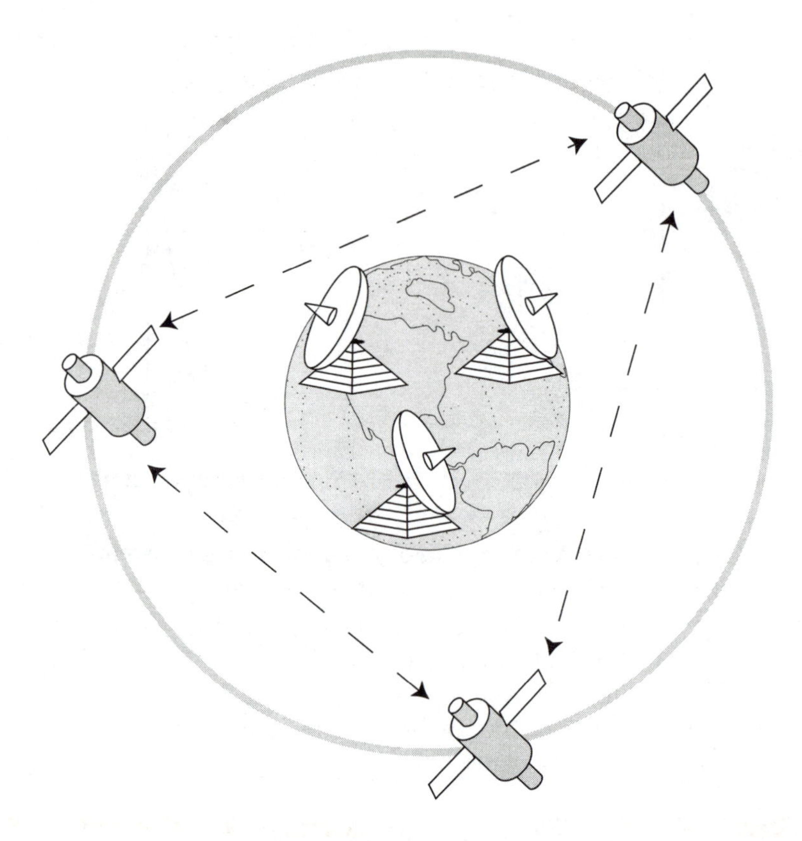

FIGURE 3-7 Geosynchronous satellites

Satellite Communications

Satellite communication is just another form of microwave transmission. Instead of aiming the signal between two or more terrestrial dishes, the signal is aimed at the satellite hovering at various orbits above the surface of the earth. While there are many orbits, geostationary or geosynchronous has been used extensively for the transmission of voice and broadcast/CATV traffic. Geostationary satellites are positioned 22,300 miles above the surface of the earth. At that altitude three to four geostationary satellites can cover the entire surface of the earth (Figure 3-7).

The most commonly used frequencies for transmission by satellites in geostationary orbit are designated as bands such as Ka, Ku, C, and L.

Table 3-1 Satellite Frequency Bands

	Frequency in GHz		
Band	Downlink	Uplink	Applications
L	1.5	1.6	Land mobile satellite Maritime mobile satellite
C	4	6	Fixed satellite data services
Ku	12	14	Fixed satellite services (voice, data, video) Radiodetermination satellite services (RDSS) and vehicle management Low-speed mobile data services
	2	18	Direct broadcast satellite (DBS)
Ka	20	30	Experimental fixed satellite services

The premium frequencies for telecommunications transmission via satellite occur in the 1–10 GHz range with data rates in excess of 1 billion bits per second (Table 3-1).

Transmission impairments in the form of attenuation, atmospheric absorption, fading, and a host of environmental and human transient impairments selectively plague satellite transmission in frequencies above and below the 10 GHz range. As a result of these engineering considerations, the C band and a significant portion of the Ku bands have become saturated.

The saturation of spectrum and economics of geosynchronous or GEO satellite transmission has intensified interest in exploring other orbits, such as medium earth orbit (MEO) and low earth orbit (LEO), for the provision of mobile satellite services. Of these two orbits, LEO represents the next frontier for mobile satellite communications services. LEO orbit is located 300 to 500 miles above the earth. As a result of this close proximity, the economics and mechanics of placing a satellite in LEO is significantly competitive with GEO satellite services.

LEO satellites are considerably smaller than GEO satellites (350 lb.), and operate with low-power requirements (97 watts). Subsequently, LEO satellites are less expensive to manufacture and emplace than its GEO counterpart. To provide total coverage of the earth's surface seventy-three LEO satellites are required. Companies such as Motorola, Elipsat,

Loral Qualcomm, Leosat, and Starsys are just a few of the companies developing LEO-based networks and services.

Ubiquitous coverage has been an elusive commodity for most cellular, paging, and mobile data communications services. The integration of these terrestrial mobile communications services with satellite transmission (LEO or GEO) could potentially satisfy the transmission piece of the coverage puzzle by providing a ubiquitous transmission platform for various mobile wireless applications.

Cellular

There are four basic components of a cellular system:

Spectrum

Mobile Telephone switching office

Cell sites

Subscriber equipment

Based on FCC frequency allocations, cellular service in the United States has been established in the 800 MHz to 900 MHz range. The range of frequencies assigned to the cellular system is divided into individual channels. Each channel has limited information-carrying capacity. The number of calls that can be handled simultaneously is determined by a complex arrangement of modulation techniques and access methods deployed and number of cells. The typical analog cellular system can support over 500,000 subscribers on approximately 832 channels that are reused many times in different cells.

Current debate over how to maximize current and future spectrum allocations has focused on different access methods. Essentially, access methods dictate how the bandwidth of a given signal or range of frequencies is used. The key concept behind access methods is providing multiple user access to limited spectrum resources. The current cellular system in the United States is analog and uses a combination of Time Division Multiple Access (TDMA) and Frequency Division Multiple Access (FDMA) scheme.

As the United States cellular industry prepares to migrate to digital cellular networks, three access methods have caught the fancy of the wireless industry. The current FM system assigns a single user to each

Table 3–2 Cellular Access Methods

Access Method	Description
FDMA	Frequency Division Multiple Access – Each user is assigned to a single frequency or signal among a band of available frequencies.
TDMA	Time Division Multiple Access – The total available bandwidth is divided into time slots. Each user is assigned a discrete time slot, permitting multiple users to share the same bandwidth sequentially. (Digital)
FDMA/TDMA	Frequency Division Multiple Access/Time Division Multiple Access – A hybrid combination of FDMA/TDMA is currently being used. The available bandwidth is divided into separate channels. Each channel is then divided into multiple time slots. (Digital)
CDMA	Code Division Multiple Access – Also referred to as spread spectrum. Code sequences are used to provide multiple channels within the same broad of channel. Each channel is designated as PN_1, PN_2, etc.

channel. If all available channels are occupied, a busy condition prevents other users from accessing the system until available system capacity is restored. Several access methods have been proposed as a way of assigning multiple users to a single signal (Table 3-2).

SPECTRUM. From 1983 to the present, rapid growth in the cellular industry has caused capacity constraints in analog cellular systems in many of the largest metropolitan areas. Overutilization, plans to migrate to digital cellular, and anticipation of new wireless services have prompted cries throughout the industry for more spectrum allocations or a repatriation of spectrum used for military and government services.

MOBILE TELEPHONE SWITCHING OFFICE (MTSO). The MTSO is the control center of the cellular network. Large intelligent switches similar in function to those used in wired networks perform a complex series of processes that coordinate call setup, routing, authentication, maintenance, and termination. In addition to cellular network operations, the MTSO is the focal point for system capacity planning, problem management, and administration, i.e., billing. Overall cellular network management includes the aforementioned functions and oversees the integration of signaling, intelligent databases, and switching activities.

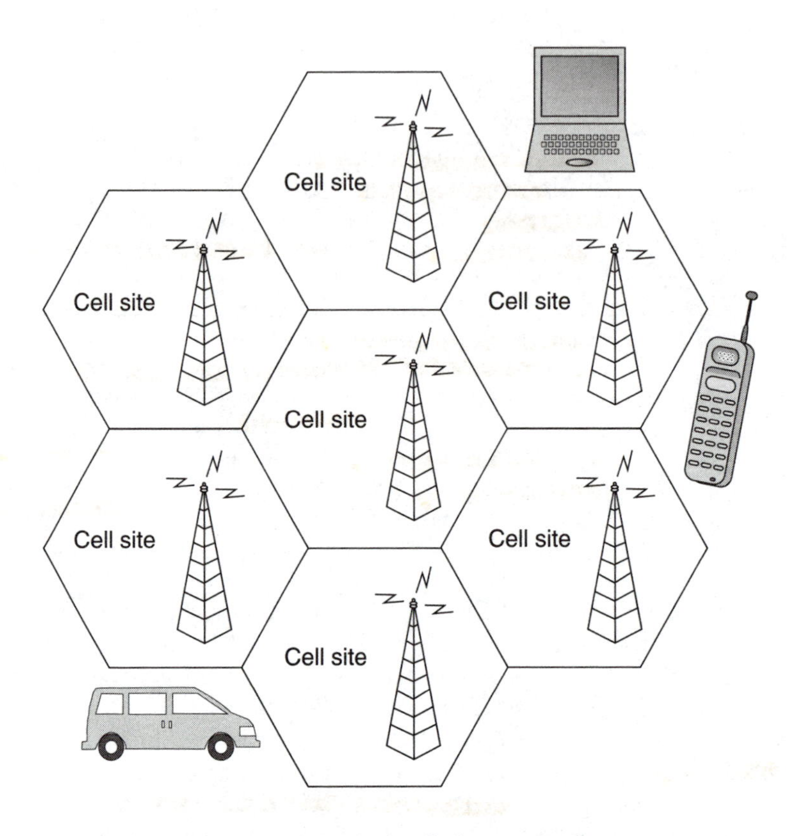

FIGURE 3-8 Typical cellular network

CELL SITES. Cellular coverage of the Cellular Service Area (CSA) is provided by a complex transmission overlay of hexagonal cells. Each cell site controls and monitors a series of frequencies allocated to provide coverage over a highly defined geographic area within the CSA. Each cell site includes an antenna, low-power transceiver, cellular control unit, call processing units, and redundant power equipment (Figure 3-8).

The size and placement of each cell within the CSA is determined by the size of the subscriber base, traffic, topography, and optimum coverage. Typically, cell site coverage ranges from 1 mile to 50 miles. Cells from several hundred feet to 1 mile in coverage are often referred to as *microcells*. Frequency reuse is achieved by assigning frequencies to cells in such a way that the same frequencies are separated by several cells; the assignments are noncontiguous.

CALL PROCESSING. Cellular call processing is a dynamic process based on frequency reuse and intelligent switching. Each cellular call is assigned to a frequency based on its proximity to the strongest available signal. As the caller moves throughout the CSA, the call signal is dynamically monitored for strength and location.

When the boundary of the active cell is approached, the cell site searches for the strongest available signal in the adjacent cell, coordinates the pending transfer with the new cell site, and notifies the cellular telephone to retune to the new signal's frequency. The process of maintaining the cellular call from cell to cell is referred to as the "handoff."

The following is a simplified flow of call processing events:

1. Subscriber unit is powered on – The activated subscriber handset is monitored by the MTSO and nearest cell site.

2. Call initiation – The caller enters the number of another cellular subscriber or site serviced by the PSTN and presses the "send" button.

3. Call setup – The number entered and information identifying the device, calling options, and location are transmitted from the cell site to the MTSO for authentication. Upon successful authentication a frequency is allocated to the subscriber unit. The call is established or the user is notified of the status of the attempted call (number not in service or busy).

4. Call maintenance – The cell site monitors the status of the call, location of the subscriber unit, and signal strength as the cellular customer moves within the cell. As the subscriber unit approaches the boundary of the cell, there is a subsequent decrease in signal strength. The active cell site initiates the handoff and releases the assigned frequency for reuse by another subscriber.

5. Call termination – When the cellular user terminates the call—presses the "end" button—all resources used for the call are

Table 3–3 Subscriber Options

Option	Description
Homer	Cellular subscriber calling behavior is generally conducted within the boundaries of the "home" system.
Roamer	Cellular subscriber calling behavior includes the "home" cellular system and frequently includes calls made while traveling in remote cellular systems.

released to the system. Information gathered by the cell site is transmitted in real time to the MTSO, which coordinates cell site activities for multiple users. This information includes origin, destination, and duration of the call, and becomes part of a database of information used to optimize system use, assist in capacity planning, and generate customer billing.

The division of the United States into CSAs and RSAs has resulted in the proliferation of separate cellular systems. CSAs are used to provide cellular coverage in metropolitan areas while RSAs (rural service areas) provide coverage in rural areas which are less densely populated. Until recently, the use of a subscriber's cellular telephone was restricted to the user's "home" system. Cellular operators around the country responded by developing "roaming" arrangements, which allow users access to service where cellular systems are available (Table 3-3). Several types of roaming behavior and services have been identified (Table 3-4).

SUBSCRIBER EQUIPMENT. While methods of classifying cellular subscriber equipment and cellular telephones vary from vendor to vendor, the telephone technology used to provide access to the cellular system and its services can be generically classified as mobile, transportable, hand portable and pocket portable (Table 3–5).

Personal Communications Systems

The document from the United Kingdom, *Phones on the Move—Personal Communications in the 1990s* is generally associated with the advent of the PCN (or PCS as it is referred to in the United States) concept in Europe.

Table 3–4 Roamer Services

Type of Service	Description
Infrequent	A temporary line dialing number (TLDN) is assigned to the visiting subscriber. The user is essentially contracting for temporary system use within the distant network.
Passport	The cellular customer's number is registered with other cellular service providers around the country facilitating unified billing.
Follow Me™	An advanced call forwarding system that allows calls made to the cellular subscriber to be forwarded when outside of the "home" calling area. The customer is billed on a time of day or peak time basis, the cost of the long distance call, and forwarding.

Table 3–5 Cellular Subscriber Equipment

Classification	Description
Mobile	Location of the telephone fixed to a moving vehicle or variable location. Power is derived from the vehicle or a separate battery pack.
Transportable	Location of the telephone is independent of a fixed location; typically carried. Units weigh several pounds and require large battery packs or interconnection to alternate sources of power.
Hand Portable	A variation of mobile and transportable cellular telephones. A single unit integrates transmitter, receiver, dialing, and power. Battery life is shorter than mobile or transportable due to the need to balance portability with size. Weight varies from vendor.
Pocket Portable	Includes all of the functions and features of hand portables. The primary differences are size and weight. Pocket portables are designed to fit easily into a shirt pocket and units are weighed in grams instead of pounds.

Personal communications (PCS) is similar to cellular service. The key differences are that personal communications services will provide digital wireless access using telephones smaller that the typical cell telephone. PCS will provide coverage much in the same as cellular through the use of cells. These cells or microcells are much smaller than those used for conventional cellular transmission.

The early specifications for PCS focused on the development of separate networks operating in the 1.7 to 2.3 GHz range to provide personal communications. The key elements of PCN include:

- Ubiquity – the ability to originate and receive calls throughout Europe
- The provision of calling capabilities independent of location
- Integrated custom calling services such as paging, conferencing, call forwarding
- The assignment of a Personal Identification Number (PIN) that will integrate all of the custom features and communications services of the user to a single telephone number.

Chapter 4

Computers, Computers, Computers

COMPUTING AT A GLANCE

Unlike other artifacts in our surroundings, computers do not lend themselves to simple explanation. When we were presented with a fork or chopsticks for the first time, we were able to directly manipulate these objects and draw some conclusions as to their probable use. When we sit at a computer for the first time, many of us stare curiously at the strange array of lights, odd shapes, and seemingly mysterious, all-powerful entity. Popular media and most people associated with computing do little to allay these fears. They seem to speak a strange dialect made up of codes and acronyms. We are made to feel like outsiders in this strange new world.

To add to this confusion there is an endless range of choices given a simple task such as buying a new computer monitor. Then there are cards for sound and communications. Do we need an 8-bit card, 16-bit card, or a 32-bit card and what does all this really mean? This chapter is devoted to unraveling the basic mystery of computer technology. It does not matter if the computer that is the object of your confusion is an IBM or Macintosh. What is important to understand is that each of these computers with different architectures still share common facilities and capabilities.

So just what constitutes a computer? And how does it work? The combination of monitor, keyboard, system unit, and other peripheral devices make up a typical computer system (Figure 4-1). On closer analysis all computer systems are made up of the following components:

- Input devices – provide an interface between the user and computer.
- Processor – is the brain or engine of the computer used to process user data and computer instructions.
- Secondary storage – provides a repository for user data and instructions.

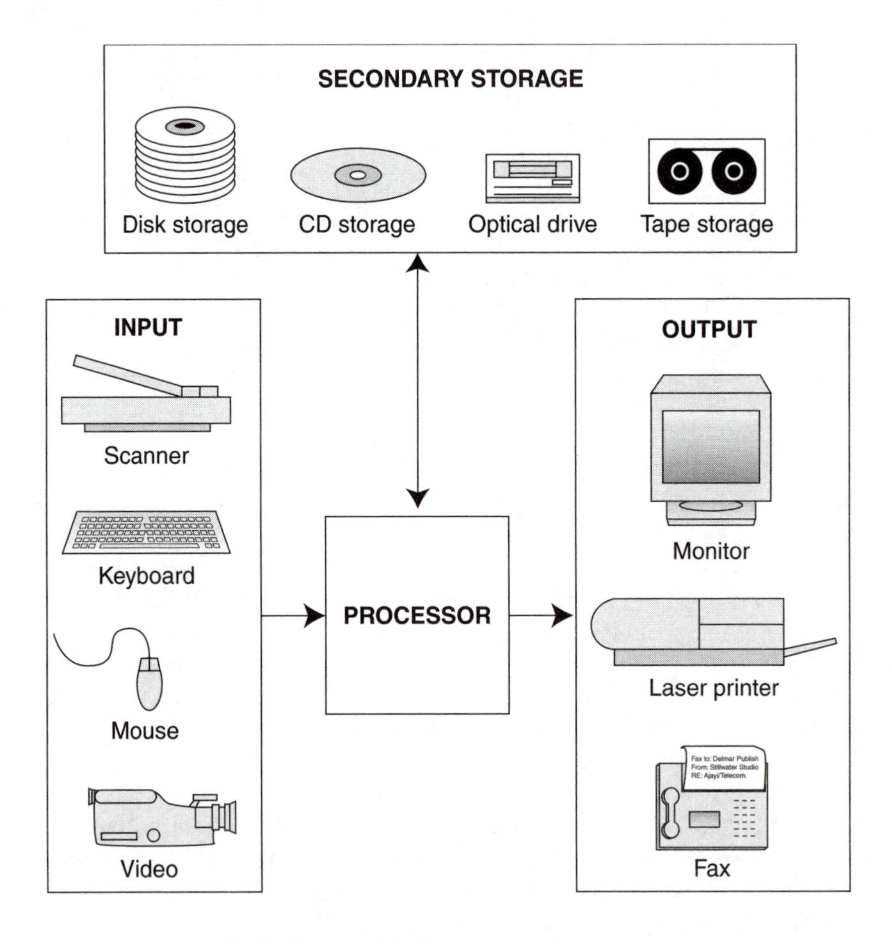

FIGURE 4-1 Modern computer system

- Output devices – provide a means for translating computer or machine language into a form that can be understood by humans and other devices.

Input Devices

There is no end to the variety of input devices available to interact with computer technology. There are keyboards, mice, joysticks, tablets, and other devices that bear no resemblance to objects we have seen before. A simple way to determine which input devices are most suitable for a specific user is to consider the age of the user, his or her respective cognitive capabilities and computer skills, the primary tasks for which the computer will be used, and any physical constraints. Once you become accustomed to the use of computer technology, you will likely switch between a number of different types of input devices. In fact, you may find that your interaction style will require the use of a combination of input technologies. You may use a mouse to perform some tasks while using the keyboard for others. In some instances, you may use speech recognition tools such as the one used to create this book.

Processor

The processor or central processing unit (CPU) along with main or primary memory form the core of modern computer systems (Figure 4-2).

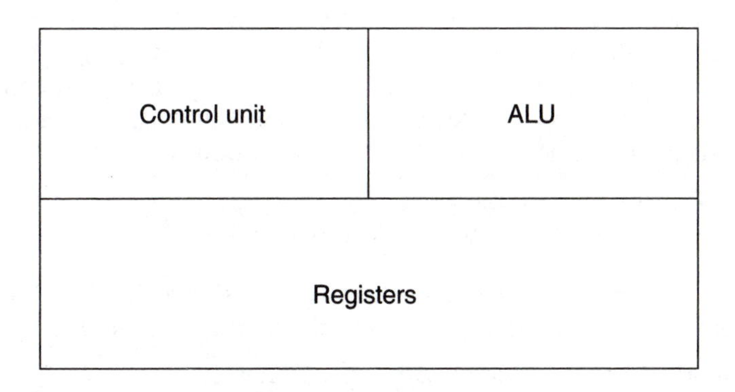

FIGURE 4-2 Central processing unit

Both IBM and Macintosh computers use CPUs that have the following components:

- Control unit – regulates the flow of data and instructions that will be processed by the CPU.
- Arithmetic logical unit or ALU – executes program instructions based on a number of logical operations such as and, or, not, less than, >, and equal to. Almost all program instructions can be executed as logical or arithmetic functions.
- Main memory – is also called *primary storage* or sometimes *registers*. This area of memory on CPU is used to temporarily store the results of the various steps used to execute program instructions. Given the limited space on the CPU, main memory is also implemented using RAM or random-access memory. Access to primary memory is coveted. It is used to store the output of processing instructions, any data that might be needed by user-application programs, portions of the operating system necessary for basic operation, and finally the user application.

In addition to the elements that make up a CPU, there are a number of other concepts that are important in understanding the basic operations of modern computer systems. In order to interact with computer technology a language or code that is acceptable must be used. Most computer systems are able to manipulate binary digits or bits represented as a 0 or a 1. These two symbols form the basis of the binary numbering system. The bit represents the smallest of elements that can be manipulated by the CPU. 8 bits are grouped into a byte or octet. Bytes form the basis of what we call code sets. The two most commonly used code sets are American Standard Code for Information Interchange (ASCII) and Extended Binary Coded Decimal Interchange Code (EBCDIC).

ASCII – is a 7- or 8-bit binary code used for transmission over communication links and by microcomputers.

EBCDIC – is an 8-bit binary code that was developed by IBM Corporation to be used primarily with mainframe computers.

The CPU processes information in a series of operations known as the *machine cycle*. The machine cycle is composed of two constituent cycles. The two cycles that make up the machine cycle are called *I-time* (instruction time) and *E-time* (execution time).

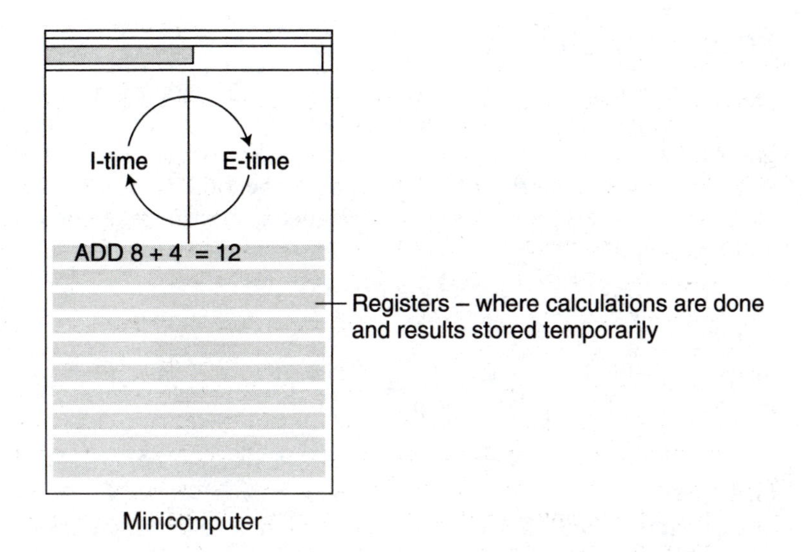

FIGURE 4-3 Machine cycle

I-time

The control unit is that element of the central processing unit that initiates the first two steps of the machine cycle (Figure 4-3). The two steps that make up I-time are

1. Fetch the next program instruction or data
2. Decode the instruction or data

E-time

The ALU accepts the decoded instructions from the control unit and performs the last two steps of the machine cycle (Figure 4-3). The steps are:

1. Executes the instruction (adds, subtracts, multiplies, divides, or compares)
2. Stores results of the process or execution in main memory or registers

Secondary Storage

Secondary storage provides a means for storing programs and user data. In addition to providing reliable high-capacity storage for data and program instructions, various forms of secondary storage are used extensively for backup and archival purposes. Secondary storage can be classified into several major types: magnetic, optical, magneto-optical, and holographic.

Magnetic

Magnetic media in the form of tapes, diskettes, and hard drives provide reliable cost-effective data storage.

Optical

Optical storage represents a major enhancement over magnetic storage media. While optical storage has been available for some time its write-once, read-many (WORM) and typically higher cost (than magnetic alternatives) kept it well out of the reach of consumer budgets. Recent increases in the amount of data that can be stored on portable magnetic media (100 MB to 10 GB) have made them a more attractive alternative to optical media. Access speed and storage capacity make optical an excellent medium for archival or program distribution. New innovations in rewritable optical secondary storage will likely replace magnetic storage as the medium of choice.

Magneto-Optical

Magneto-optical is a hybrid combination of magnetic and optical storage technology. The ill-fated floptical is an example of magneto-optical storage technology.

Holographic

Holographic is the real sci-fi end of the secondary storage spectrum. Holography is being explored as a medium for the next generation of computer technology secondary storage and random-access memory.

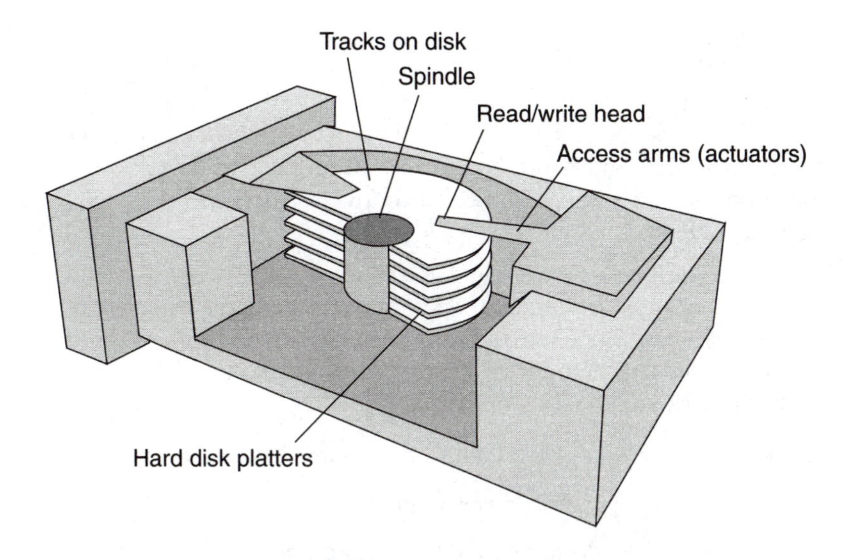

FIGURE 4-4 Hard drive cutaway

Hard Drives

The addition of hard drives to microcomputers ushered in the age of personal computing. In addition to providing a storage medium for the operating system, application software, and user data, large capacity hard drives have made interactive multimedia a reality. The hard drive is organized into cylinders, tracks, and sectors (Figure 4-4). The speed at which a given hard drive is able to transfer data is equally as important as its capacity. In order for data and instructions to be transferred from a hard drive, four elements must be completed. These four processes are collectively referred to as *access time.*

1. Seek time – the time it takes to find the appropriate track on the hard drive

2. Rotational delay – the times it takes for the read/write heads to be positioned over the desired data and instructions

3. Head switching – the selection between the read head when transferring data from the hard drive, or write head for transferring data to the hard drive

4. Data transfer – movement of data and instructions to and from the hard drive

Rotation speed, which is measured in rpm or revolutions per minute, is a major factor contributing to higher transfer rates of data and instructions from the hard drive. Typical hard drive rotation speeds range from 4,000 to 7,200 rpm.

Tracks and sectors – are the surfaces upon which user data and information are written on the hard drive. The outer segments of the disk have more sectors than the inner section of the disk. In other words, data and instructions that are stored toward the end of the disk are accessed more quickly.

cylinders – all of the tracks on the same disk surface

tracks – concentric circles on a cylinder

sectors – tracks are divided into fixed length segments made up of blocks that contain data and instructions

Disk access time – is a combination of seek time, rotational delay, head switching time, and data transfer. Data access time is measured in milliseconds.

IDE OR SCSI. There are two popular formats for hard drives – Integrated Drive Electronics (IDE) and Small Computer System Interface (SCSI). In the past, IDE drives were used in WinTel machines and SCSI was used with Macs. In addition to speed and format, these drives differ in terms of the requirement for an external I/O controller card or whether the I/O circuitry is integrated within the drive.

IDE – Integrated Drive Electronics is a type of hard drive that contains a built-in controller. The IDE drive is connected to the IDE host adapter with a flat ribbon cable. High-speed integrated host adapters can support multiple floppy drives, multiple hard drives, and other types of IDE input/output devices.

SCSI – Small Computer System Interface offers flexibility in microcomputer input/output options. By using a SCSI card, up to seven devices can be interconnected using a single computer expansion slot. SCSI devices such as Iomega's JAZ and ZIP drives, hard drives, digital audio tape (DAT), and CD-ROM can be added to existing computer configurations with ease. There are several standards specified for SCSI. SCSI-1 is first generation 8-bit SCSI technology that supported transfer rates of up to 4 MB. SCSI-2 provides 10 MB transfer rate using 16-bit architecture. More recently, wide SCSI or ultrawide SCSI-2 provides both 16- and 32-bit support with data transfer rates of 20 MB and 40 MB per second respectively.

Other Secondary Storage Technologies

CD-ROM and other portable mass storage media have undergone a number of significant changes. Bernoulli and SyQuest drives were popular within the graphic arts and printing environments, but suffered from a generalized lack of availability outside of these specialized environments. Iomega introduced a number of portable storage media that capitalizes on the strengths of earlier storage technologies such as Winchester, Bernoulli, and SyQuest. Zip drives boast a portable storage capacity between 100 MB and 1 GB for a few cents per megabyte.

Compact disk read-only memory (CD-ROM) and its sister technologies offer the benefits of mass storage (600 MB+), portability, and competitive pricing. CD-ROMs are rapidly replacing floppy disks as the medium of choice for software distribution. 24X speed and higher CD-ROM drives provide high-speed access to photos, images, and software applications.

CD-R or Compact Disk-Recordable drives allow users to "burn" their own disks for in-house or external use. Adding record capability made CD-R an attractive optical storage alternative. However, many users turned to high-capacity rewriteable magnetic portable media because first generation CD-R technology was write-once, read-many (WORM). The industry responded by introducing rewriteable CD-ROMs called Compact disc-Rewriteable (CD-RW).

DAT. DAT or digital audio tape represents an enhancement in sequential magnetic media. Used as the standard for professional recording, DAT never took off in the consumer market. By using a helical arrangement or digital data storage (DDS) the current generation of DAT technology can hold between 2 and 24 GB of computer data. Using DDS-3, DATs can be formatted to hold as much as 24 GB of data with transfer rates up to 2 MB per second.

DAT cartridges come in a number of formats: 1/2 inch, quarter-inch cartridge (QIC), 8 mm and 4 mm. The following table shows the formats and formatted capacities of DAT cartridges.

Format	Formatted Capacity
1/2 inch	60 – 400 MB
QIC	40 MB – 4 GB
8 mm Helical	1 – 4 GB
4 mm DAT	1.3 GB

DVD. Capacity-hungry interactive multimedia applications continue to challenge the upper limits of magnetic media. Until recently, optical storage technologies were plagued by high costs and its WORM capabilities. The industry has responded by developing a whole new class of rewriteable media. The goals driving advancements in optical storage technology include gigabyte storage capacity, rewrite capability, and platform integration. The primary goal of platform integration is to develop optical technologies capable of storing and outputting voice, images, video, and other forms of interactive multimedia traffic on or from a single medium. For example, a typical interactive multimedia environment might include a CD-ROM or CD-R, a videocassette recorder and camera, a video capture board, and other specialized devices in order to accommodate a potential range of multimedia traffic.

DVD or digital video disk (more recently dubbed digital versatile disk) is a new optical disk technology that many predict will replace magnetic and optical alternatives such as VHS VCRs, CD-R, CD-ROM, and other forms of portable storage media. When compared to existing technologies such as CD-ROM, DVD not only offers increased storage capacity (17 GB) but it offers record capability on both sides of the medium. DVD standard supports backward compatibility with CDI, photo CD, CD-R, CD-ROM, and other hybrid CD-based technologies.

Various versions of DVD technology are starting to appear in consumer electronic outlets. New multimedia workstations are being shipped with DVD as a standard replacement for CD-ROM. DVD players are predicted to replace VHS VCRs in three to five years.

Output Devices

Output devices act as an interface between the user and computer by providing a means of displaying or printing information or data. Video display terminals (VDTs) along with high-quality desktop printers form the bulk of present output alternatives. While there have been some noteworthy developments in audio output in the form of speech-to-text and the addition of voice files to e-mail, audio applications offer limited functionality while primarily providing playback and alerting functions.

SOFTWARE

In order to exploit the tremendous power of modern-day computers, software in the form of operating systems and application programs are necessary. Over the past decade, the issue of which computer platform (the combination of hardware and software) is most suited to consumer and corporate needs has nearly been laid to rest. The introduction of graphical user interface (GUI) has leveled the playing field between WinTel and Macintosh computers in terms of usability.

Operating Systems

Operating systems provide the essential control, coordination, and monitoring functions for computer systems. While there are many types of operating systems, environments, and shells, these software resources can be grouped on the basis of platform dependence or independence—portable and proprietary. Portable operating systems are those that are platform independent. UNIX is an excellent example of portable operating system, which can be run on a diverse range of computing platforms. While UNIX offers the user powerful tools, its lack of a user-friendly interface has kept it from gaining popular acceptance outside of programming environments.

Proprietary operating systems are platform specific and will only work on computer systems and with applications software that is specifically designed for it. A number of proprietary operating systems have emerged as a result of the fierce competition for the modern desktop. It is ironic to note that many of the key players in the operating system environment such as Microsoft, IBM, and Macintosh once worked together as strategic partners. The following is a brief list of popular proprietary operating systems available today.

Windows 3.1 – graphical shell

Windows NT – 32-bit network operating system

Windows 95 – 32-bit operating system

Macintosh OS – for Apple computers

OS/2 – IBM's 32-bit offering

Applications Software

The advent of microcomputers stimulated the widespread availability of off-the-shelf application programs. The days of in-house programming were soon replaced with a rich variety of programs such as word processing, graphics and image, database, and spreadsheet. More recently applications software has been bundled in the form of productivity suites. These application suites often combine a powerful word processor, graphics, spreadsheet, database, and communications functions into a single integrated offering. The chief benefit of these productivity suites lies in the integration and portability of data and information between the various components. For example, data contained in a spreadsheet can be turned into eye-catching graphics with a click of a button.

IN SEARCH OF SPEED

In addition to the high demands placed on storage and access to information, interactive multimedia requires robustness from the platform upon which it is built. Nearly a decade ago, the typical computer configuration boasted random-access memory of 412,000 bytes, 20-MB hard drives, and 13-inch monochrome displays. Today's configurations are characterized by the 32-bit and 64-bit add-on video, sound, I/O, and other special function cards. Memory requirements for these configurations have climbed upward of 128 MB of RAM and secondary storage options, which include both optical and magnetic alternatives in the multi-gigabyte range.

When confronted with the task of buying a new computer, many users feel as if they are stepping into the eye of a hurricane. The following are some terms and definitions that should be understood in determining the factors that contribute to optimum system speed.

Clock speed – a metric of the number of pulses generated from a quartz crystal; used to provide timing for the CPU. CPU speed is typically expressed in MHz; i.e., 300 MHz Pentium II processor.

Clock – refers to the internal or system clocks used in the computer environment. The CPU clock uses a quartz crystal to synchronize the processing of instructions per some given time interval.

Clock doubling – technique used by experienced computer users where parameters changed in the configuration setup contained in the Complementary Metal-Oxide Semiconductor (CMOS).

Clock pulse – refers to the signal generated by the internal CPU clock. These pulses are continuous and correspond precisely to changes in voltage.

Bus width – refers to the size of the electrical pathways that connect the various subsystems of the computer to the CPU. Bus width is typically expressed 8-bit, 16-bit, 32-bit, and 64-bit. With a larger bus width, larger or more instructions and data can be sent to the CPU for processing.

MIPS – millions of instructions per second; perhaps a more meaningful metric of system speed for personal computers.

Megaflops – in minicomputers and mainframes megaflops refers to 1 million floating point operations per second.

Determining the exact combination of factors that contribute to speed is far from an exact science. Popular computer literature is flooded with confusing benchmarks that suggest that the clock speed alone is not a sufficient metric of performance across a range of applications. To add to the confusion, the presence of clone or hybrid processors tout processing speeds faster than native Intel or Motorola CPUs.

MAKING SENSE OF MEMORY

The concepts of memory are perhaps the most misunderstood among all factors associated with computers. There are RAM, ROM, SIMM, DIMM, EPROM, and a host of equally confusing acronyms. The goal of memory is to provide a workspace for the execution and manipulation of computer instructions and user data. Perhaps the most important type of memory is referred to as *primary memory, memory,* or *main memory.* On the CPU, some main memory exists in the form of registers. Registers are temporary holding areas for the results of computations performed by the ALU. Given the limited amount space available on CPU, RAM offers an extension of this coveted workspace.

So just what is RAM, cache, and all of the other terms? Let us start with a generic overview of memory. As stated earlier, memory is the workspace for the computer. Memory is addressable. The analogy is similar to the main post office where P. O. boxes are located. Each box has a discrete number associated with it. As instructions and data are read into and out of memory, they are stored randomly on the hard drive or other secondary storage media. The location of data and instructions is maintained by a table that is analogous to the yellow pages. So why are there so many types of memory?

Perhaps an easy way to approach the topic of memory is to look at the ways in which the content of memory is retained. Two categories of memory exists:

1. Volatile – electricity must be supplied to this type of memory to retain data and instructions. Static random access memory (SRAM) and dynamic random access memory (DRAM) are types of memory chips that require power to retain their contents. SRAM is typically faster than (10 ms) DRAM; however it is more expensive. Consequently, DRAM is used in the majority of computer configurations (60–70 ms).

2. Non-volatile – memory that falls into this category retains its contents even after the power is turned off. Typically, default settings for hardware and other firmware components are "burned in" at the factory. PROM or programmable read-only memory is another type of non-volatile memory. With PROM the user can program or alter the contents of the chip.

Cache memory is another type of dedicated high-speed memory used to hold the most frequently used data and instructions. The benefit of cache is that it attempts to predict which data and instructions will be needed. If the data or instructions are not in cache or main memory, then they must be located and retrieved from the hard drive or other medium, slowing program execution.

CLASSES OF COMPUTERS

Now that we have explored the intricacies of computer technology, we can look at the broader categories into which computers are divided. Three types of computers will be discussed in this section: mainframes, minicomputers, and microcomputers.

Mainframes

The earliest computer configurations were called *mainframes*. These were typically large footprint computers that supported centralized data processing. First generation mainframes did not lend themselves to a high degree of user interaction. Those individuals who interacted with mainframe computers were highly trained programmers and other computer professionals. There are a number of benefits associated with mainframe computing. Mainframes are large, robust machines capable of processing millions of floating point operations per second. Since all computer operations were centralized, standardization was easy to maintain. There were, however, a number of limitations associated with mainframe computing. First, the sheer cost of these environments meant that only the largest organizations were able to afford them. The centralization of processing in these environments represented an increased risk in terms of reliability and disaster recovery. Finally, most applications in the early mainframe environment required in-house programming.

Minicomputer

Enhancements in the very large scale integration or VLSI ushered in a new era of computing. Small footprint, highly specialized computers known as *minis* or *minicomputers* often rivaling the processing power of mainframe computers began to appear. Perhaps the biggest difference lies in the diffusion of minicomputers throughout different segments of an organization. For example, the flexibility in application programming that was not available in the mainframe environment could easily be provided to functional areas that had shared data requirements by using minis.

A single minicomputer could be used by accounts payable, accounts receivable, purchasing, and the sales department to gain real-time access to data. With up-to-date information on the financial status of the company and its accounts, cash flow is greatly improved within the organization. The chief benefits of minicomputers lie in the parallel development of distributed data processing and peer-to-peer networking.

By distributing computer resources in this fashion, the potential risks associated with a single point of failure are minimized. In short, the failed computer could be bypassed and its workload distributed among the remaining computers. Minicomputers, as part of network configurations, represented increased operational costs as well as increased network management requirements.

Microcomputers

Microcomputers took corporate and consumer markets by storm. The term *personal computer* became closely associated with this type of configuration because for the first time, the user could actually control and customize most aspects of their computing environment. The proliferation of microcomputers throughout the corporate world soon highlighted the need for standardization, interoperability, and increased network security. In addition to the individual power that microcomputers offered its users, it was no longer necessary to program specialized applications to meet their needs. A rich array of off-the-shelf application programs became readily available. This allowed organizations to increase productivity by giving workers the tools needed to get the job done. New network models began to emerge such as local area networking and more recently client/server. Perhaps the chief limitation of microcomputing lies in its rapid obsolescence.

Supercomputers and Workstations

Finally, there are several other categories of computer technology that are currently recognized. Supercomputers often support multiple CPUs and provide parallel processing capabilities. In this instance, multiple programs and instructions can be executed in parallel.

Powerful desktop configurations called *workstations* are hybrid microcomputers designed to perform specialized tasks. In terms of processing capabilities, many workstations are similar to mainframes and minicomputers.

Chapter 5

Network Basics

OVERVIEW

In previous sections we explored the hardware, software, and transmission elements that comprise modern networks. With so many alternatives to choose from and the need to keep information resources current, many companies have opted to outsource responsibility for information systems.

There are many ways to categorize a network. Networks can be classified by the geography covered, topology or arrangement of the infrastructure, the type of traffic that is carried, and many other factors. Let us start by defining some basic network terms.

Network – refers to the integrated collection of hardware, software, and telecommunication facilities used to connect geographically dispersed resources.

Infrastructure – refers to the complex mix of wired and wireless facilities used as the backbone or transport mechanism for the network.

Node – refers to switching, routing, and other intelligent devices that make up a network. These devices may include computers, workstations, and specialized telecommunication devices such as front-end processors, modems, multiplexers, and other intelligent networking devices.

Private network – networks typically maintained by corporate, academic, or other privately held entities. These networks may be circuit switched, packet switched, client/server, peer-to-peer, and other current networking configurations.

PSTN – the public switched telephone network is made up of telecommunication facilities provided by common carriers. The combination of local and long distance service (LDS) collectively make up the infrastructure of our public network.

Switching – in this context, refers to routing decisions that are made by intelligent nodes within the network (Table 5-1). The PSTN uses circuit switching. Networks such as Dow Jones use packet switching. E-mail is a form of message switching.

Interoperability – refers to the ability to connect to, and the degree of compatibility necessary to support the exchange of information across networks of different architectures.

Now that we have outlined a few basic networking terms, let us take a look at networks more closely. Based on geography we can identify three types of network configurations

1. Wide area networks or WAN
2. Metropolitan area network or MAN
3. Local area network or LAN

Wide Area Networks

Wide area networks typically span a city, county, region, territory, or nation. The public switched telephone network and the Internet are examples of wide area networks. WANs can be made up of voice traffic, data traffic, video traffic, or a hybrid combination of these types of data as interactive multimedia traffic. Wide area networks typically include a number of areas referred to as *domains*. Each domain consists of all the hardware, software, nodes, and telecommunications facilities within its scope. Typically, issues such as performance, capacity planning, and network security are coordinated at the domain level. Most modern WANs must support not only a diverse range of data types, but they must also terminate or carry traffic across both wired and wireless facilities.

Table 5–1 Network Description – Switching Technique	
Circuit Switched Network	Optimized for voice but provides data connectivity via modems and other devices: PSTN, PBX, KTS
Packet Switched Network	X.25 network that breaks data into information packets for transmission through the network. Store and forward/real time: PDN
Message Switched Network	Messages are sent (data or voice) as complete units through the network. Typically not real time. Store and forward: E-mail, voice mail

Metropolitan Area Networks

Metropolitan area networks are the type of network that is specified by the Institute of Electrical and Electronic Engineers (IEEE) standards for local area networking. MANs typically cover a few city blocks to a few miles. For example, many universities are on campuses where academic and administrative offices are spread across a large territory or city. The school of dentistry, the main hospital, and academic facilities are interconnected using private, public, or a combination of these two facilities. Another example of where MANs can be deployed is with an automotive dealership whose locations are spread within a few miles of each other.

Local Area Networks

Local area networks are also specified by 802 IEEE standards. Since local area networks make up a major portion of most public and private networks, most of the attention in this section will be focused on them. In the late 1980s, the proliferation of microcomputers presented many organizations with a problem. How could they seamlessly exchange data between standalone devices that often had different platforms, protocols, and capabilities. The factors that drove the development of local area networking are:

- The need for flexible communications and information—project sensitive
- The need to move or make changes

- Structural issues—real estate, utilities
- Connectivity cabling
- The need to interface with different, proprietary applications

For example, if you manage a small company that employs 100 people, in a standalone configuration each user would have his or her own monitor, system unit, and associated devices. In addition to the investment in hardware, additional funds would be spent to provide each station with the software required to do their respective jobs. Current innovations in networking could be applied to this scenario to minimize duplication of resources, multiple points of failure, and as a buffer to obsolescence.

Advances in secondary storage, increased price performance of computers, the decline in cabling costs, and the need to share resources defined a new networking paradigm called *local networking*. In the previous example where there were 100 people employed, instead of having 100 separate workstations with associated hardware, software, and printers, we could connect the stations and a larger network printer together with the following four elements:

1. File server. The file server is a dedicated workstation whose responsibility is to manage shared operating functions and application software among connected entities. In addition to managing shared software resources, the file server also mediates access to physical transmission media and any shared peripheral devices attached to the network.

2. Network operating system. A specialized operating system (OS) is required to implement the functions of a standard local area network. These functions include routing, addressing schemes, media access, and other specialized functions necessary to support a given environment.

3. Infrastructure. In order to connect all the resources to the LAN, some form of wiring or cable is necessary. From previous sections, we know that these facilities can be wired such as copper twisted pair, coaxial cable, or fiber-optic cable. They may also be attached using wireless facilities such as laser, LED, or Infrared Data Association (IRDA).

4. NIC. In order to attach the various nodes and stations to the LAN, a network interface card or NIC is required. The network interface card may attach a computer or printer to the LAN physically when guided facilities are used. In the instance where wireless facilities are used, network interface cards will have an antenna or equivalent sensor to receive the RF signals.

UNDERSTANDING LOCAL AREA NETWORKING

Let us start with some land basics. We know that we can use wired or wireless facilities as a backbone for local network. We know that a specialized network operating system (NOS) resides in the file server to provide network management functions. Finally, network interface cards provide connectivity between user stations and the physical network. But as with any shared arrangement, there are often bottlenecks or contention for the same network resources. In order to control transmission within a local area network environment, several access methods have been adopted:

- CSMA – carrier sense multiple access is a type of LAN access method where permission to transmit is granted to a station after the medium is checked to see if it is clear.
- CSMA/CD – stands for carrier sense multiple access with collision detection. As with CSMA the network is checked to see if it is clear. If two devices attempt to access the LAN at the same time, this access method will detect the collision and adjust subsequent attempts by these devices so as to minimize another potential collision.
- CSMA/CA – stands for carrier sense multiple access with collision avoidance. In addition to incorporating the features found in CSMA/CD this access method listens before, during, and after the attempted transmission.
- Token passing – another way to provide access to network resources would be to use a specialized series of bits called a *token*. Devices wishing to transmit must first seize the token before beginning information transfer.

Topologies

The topology of a given network refers to the physical or geographic arrangement of its nodes and other resources. Differences in topology determine the ease with which moves, acts, and changes (MACs) are made to the LAN. The topology also contributes to the reliability and recoverability due to network outages or congested resources. Here are a few of the most common topologies used in local area networking:

- Bus – stations or nodes are attached to the network in a linear fashion (Figure 5-1).
- Ring – stations or nodes are attached to the network in a circular fashion (Figure 5-2).
- Star – stations and other devices are attached to the network where the file server may serve as the central point and all nodes and devices are attached to it independently (Figure 5-3).
- Tree – stations are attached to the network in a perpendicular fashion. In short, several buses are arranged in a hierarchical fashion.
- Hybrid or mesh – by combining bus, ring, star, and tree topologies within the same network, robustness and flexibility are easily ensured.

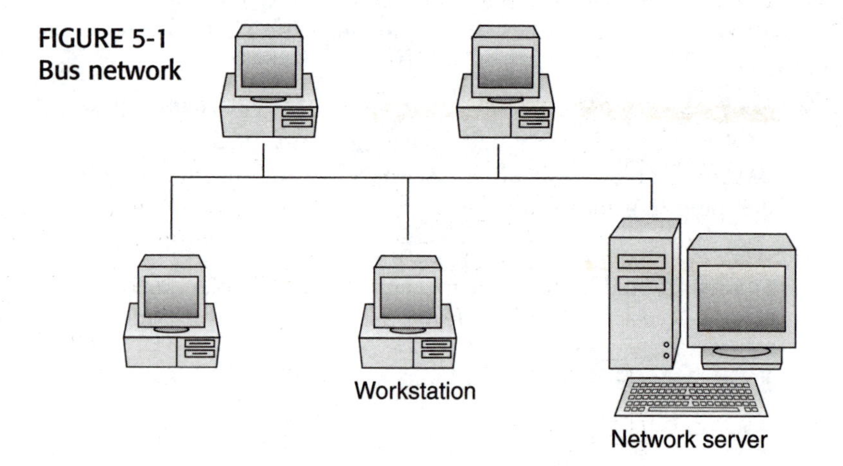

FIGURE 5-1
Bus network

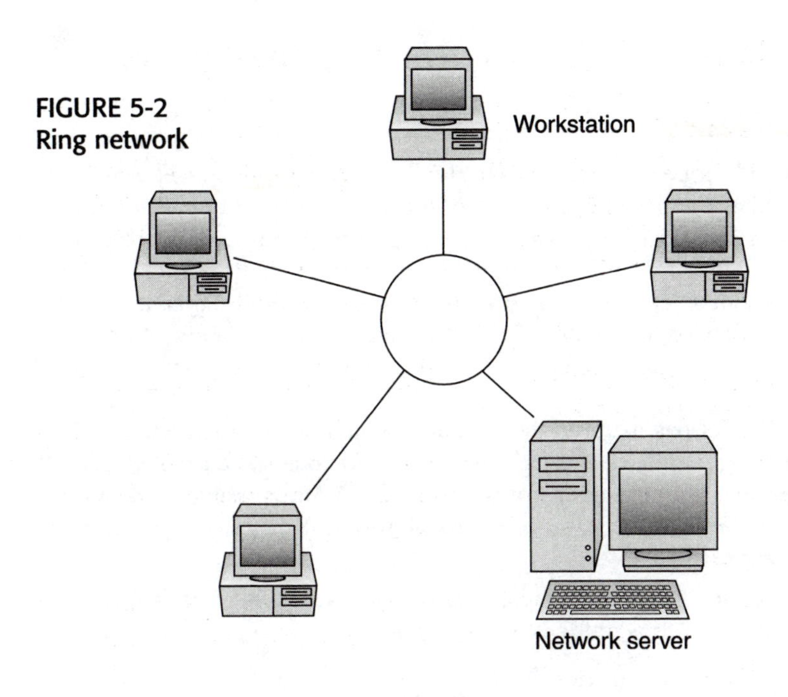

FIGURE 5-2
Ring network

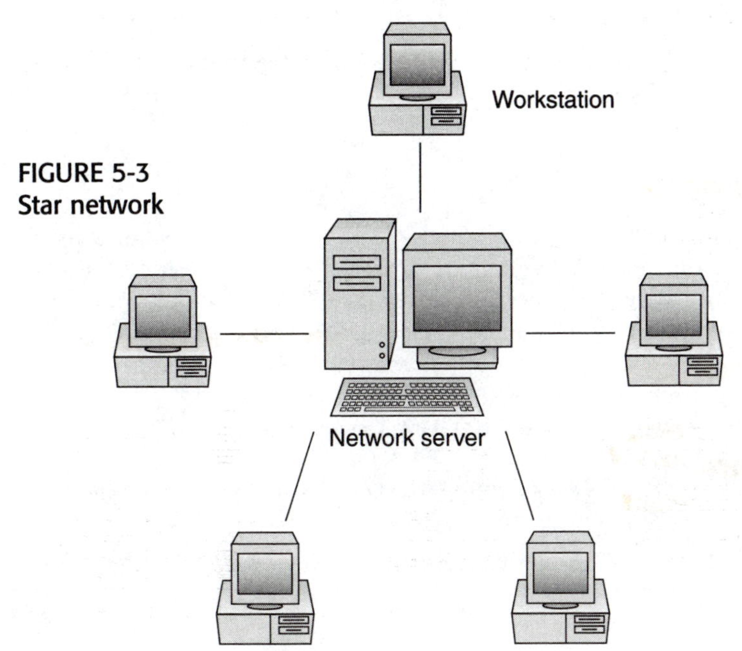

FIGURE 5-3
Star network

POPULAR LAN STANDARDS AND OFFERINGS

EtherNet

Jointly developed by DEC Corp., Intel, and Xerox Corp., EtherNet has emerged as the premier standard for local area networking. Based on the IEEE 802.3 standard for local networking, EtherNet is capable of providing data transmission rates of up to 10 Mbps to as many as 1,024 attached stations. EtherNet uses CSMA/CD as its access method. EtherNet's popularity is perhaps attributable to its flexibility across a variety of transmission media such as wireless, fiber-optic, copper twisted pair, and coaxial cable.

EtherNet has matured into a multi-megabit local area network offering. First generation EtherNet was referred to as 10Base-T. 10Base-T offered users transmission rates of up to 10 Mbps using CSMA/CD as transport protocol. Various iterations of 10Base-X have emerged and are summarized as follows:

- 10Base-2 – specifies the use of coaxial cable where each span supports 10 Mbps traffic of up to 185 meters.
- 10Base- 5 – specifies the use of a heavier gauge coaxial cable where each span supports 10 Mbps traffic of up to 500 meters.
- 10Base-F specifies the use of fiber-optic cable to provide 10 Mbps data transmission of up to 2,500 meters.

100Base-T

As local area networks became the focal point for interactive multimedia traffic, the demand for bandwidth increased well beyond the 10 Mbps that was available with EtherNet. By using more efficient network access protocols and faster transmission facilities, 100Base-T also known as fast EtherNet provides user access to network facilities of 100 Mbps.

There are three types of 100Base-T:

100Base-T4 – provides 100 Mbps over four pairs of unshielded twisted pair (UTP).

100Base-TX – provides 100 Mbps data transmission over data grade twisted pair.

100Base-FX – provides 100 Mbps data transmission over two-strand fiber-optical cable.

FDDI

FDDI or as it is pronounced FID-DEE stands for fiber distributed data interface. FDDI is based on the American National Standards Institute (ANSI) X3-T9 standard. FDDI is implemented in two versions. A fiber-optic version provides 100 Mbps transfer of data up to 124 miles. Using copper twisted pair 100 Mbps transmission is provided as CDDI. CDDI stands for copper distributed data interface; in this case, FDDI is implemented using UTP instead of fiber-optical cable.

One of the key goals of ANSI's X3-T9 subcommittee is to provide upward compatibility between FDDI/CDDI and emerging transport technologies such as SONET and ATM. Second generation or FDDI-II adds voice support to this high-speed local area network offering.

Gigabit LAN

Transmission rates close to one billion bps over local area network technology is referred to as gigabit LAN. The collaborative efforts of industry heavyweights such as Texas Instruments, Sun Microsystems, AT&T, Compaq, and 3Com have fostered the development of first generation gigabit LAN offerings capable of being implemented over a variety of guided transmission facilities. Many industry watchers speculate that gigabit LAN technology will occupy the niche market where ATM proves to be too expensive as an interactive multimedia transport alternative.

NETWORK COMPUTING —A CHANGING PARADIGM

Discussions of networking evolution and alternatives are often done in isolation of the very technology that stimulated its growth. Were it not for the widespread deployment of computer technology, many of the networking models that have become so familiar such as peer-to-peer, client/server, and Intranet, would not be possible. The ability to expand corporate information resources across a vast geographic territory has stimulated strategic expansion into virtual markets.

In order to provide a better understanding of the interaction between telecommunications and computer technology, the charts on page 76 and 77 outline three types of computers and the network opportunities and challenges they represent.

Processor	Networking Issues
Mainframe	Earliest configuration. Centralized. All devices were directly attached to the CPU. With the advent of switched lines, users could dial into the host on an as-need basis. Modems, front end processor and communications controllers were deployed.
Minicomputer	The decline in the size and availability of computer technology fostered the development of departmental or distributed processing. Since networking trends parallel those of computing trends associated transmission facilities were spread across wider geographic areas.
Microcomputer	The microcomputer revolution took corporate, small business, and consumer markets by storm. This was the first time that computer technology had the possibility of reaching saturation points similar to telephone, television, and radio.

By examining the evolution of computer technology, we can draw parallels in terms of corresponding increase in network computing. Early mainframe environments were characterized by centralized data processing where all devices were directly channel-attached to the CPU. The introduction of switched and dedicated lines to mainframe computer facilities extended computing resources among geographically dispersed locations. In order to off-load the communication functions from the mainframe itself, specialized devices such as front-end processors or communication controllers were deployed. Perhaps the chief advantage of mainframe network computing was the ease of implementation and upgrade. Given the raw processing power found in mainframe environments, software applications were more sophisticated and robust. The downside of master/slave networking is that they were typically more expensive to maintain and often were susceptible to outages due to the single point of failure.

Minicomputers, along with advances in shared database technology, were a major catalyst in the evolution of distributed data processing. Minicomputers offered smaller, more reliable devices that easily dispersed throughout the network. The benefit of DDP came in the form of availability and reliability because much of the data and processing could be switched between the networked entities. Perhaps the chief disadvan-

Processor	Advantages	Disadvantages
Mainframe	Ease of implementation and upgrade More sophisticated application resources	Expensive to maintain—equipment, software, and human resources Lack of variety in applications; development backlogs
Minicomputer	Smaller, more reliable devices. Flexible networking and intelligence. Specialization of resources to meet needs.	Network management and control burden increased as each location has to assume this role. Cost – duplication of applications, staff, and network resources.
Microcomputer	Flexibility Cost – off-the-shelf packages	Incompatibility Networking nightmare

tage associated with distributed processing lies in the increased network management burden to the diffusion of network control. In many cases, costs associated with distributed processing outstripped existing network budgets. This is due primarily to duplication of human, computer, and telecommunication resources.

The overlap between the development of minicomputers and microcomputers gave way to new networking paradigms. More efficient network protocols and the use of higher-speed digital transmission facilities set the stage for local area, peer-to-peer, and client/server computing.

Master/Slave

In the master/slave network computing model, a mainframe or powerful minicomputer acts as the primary computing resource. Other computing resources are configured as part of this network; however, they serve a secondary role in the overall processing power of the network. Master/slave computing did not offer its users much in the way of flexibility or protection from vulnerability in the form of network outages. If the primary or master computer resource became unavailable or incapacitated, computing would grind to a halt (Figure 5-4).

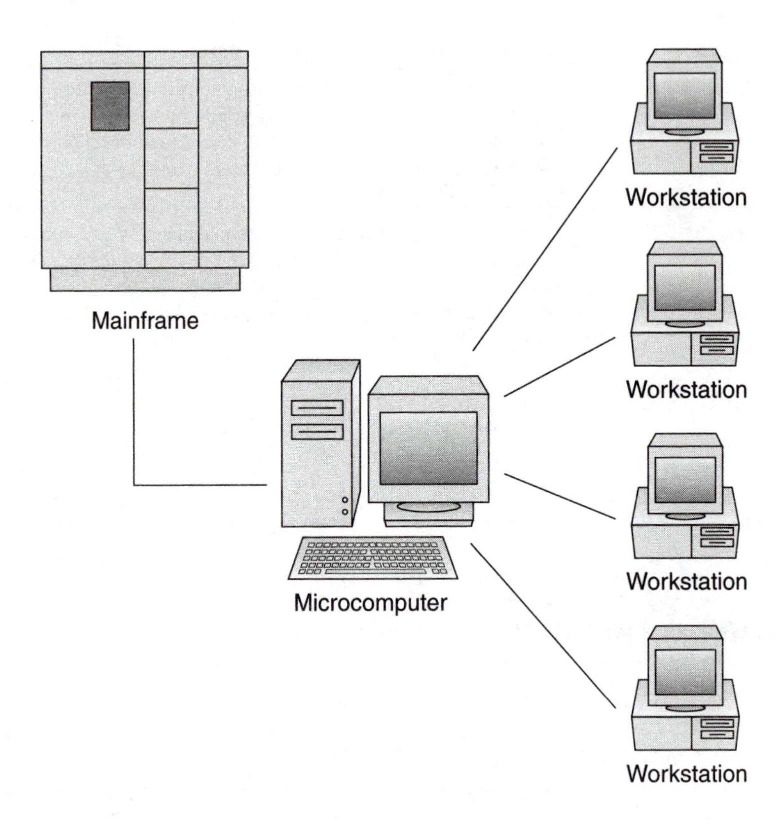

FIGURE 5-4 Master/slave network

Peer-to-Peer

Peer-to-peer networking evolved solving many of the frailties associated with the master/slave model of computing. By distributing both computing and data resources among geographically dispersed nodes, a high degree of availability and recoverability characteristic of networks built on this computing model. In peer-to-peer computing, all attached resources are capable of performing the same functions (Figure 5-5).

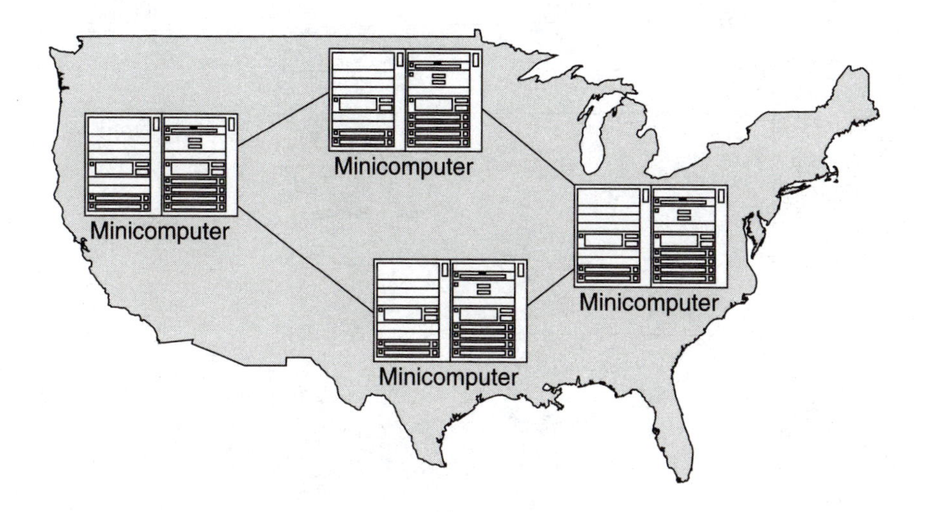

FIGURE 5-5 Peer-to-peer network

Client/Server

Recent interest in the Internet has stimulated attention in, and the growth of, client/server-based networks. In order to understand the sophistication of this networking model, the terms *client* and *server* are defined.

> *Client* – in this network arrangement, the client is that network entity that requests computing services from other servers in the network. In short, the client requests applications and utilities from other network-based software resources.
>
> *Server* – many definitions have been associated with the term "server." Within this context, server refers to a computer program that provides services on request to attached or connected clients.

The Internet using the hypertext transport protocol (HTTP:) is perhaps the most well-known client/server environment. When we log onto the World Wide Web via the Internet using a browser, the browser in effect becomes the client. As we surf the Net, we are served HTML-based Web pages by the various servers that reside on the Internet (Figure 5-6).

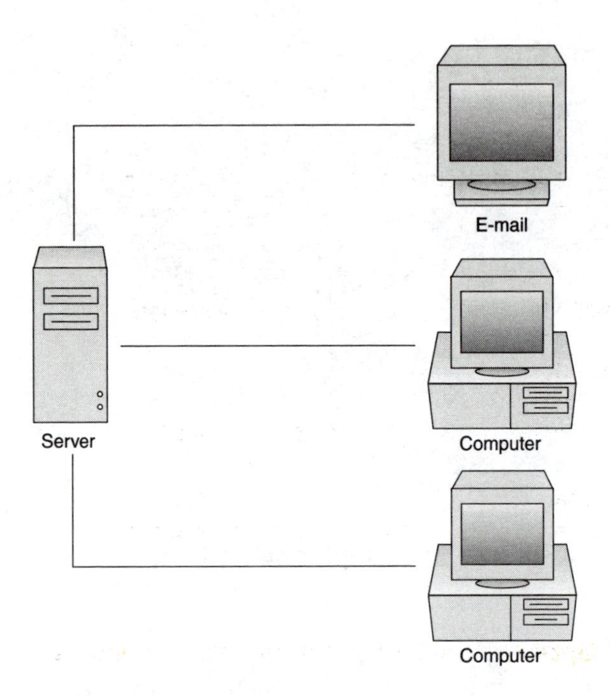

FIGURE 5-6 Client/server network

NETWORK INTEROPERABILITY

As high-speed digital networks began to connect remote parts of the globe, the issue of interoperability becomes paramount. Many companies are expanding into international markets by deploying WANs carrying interactive multimedia traffic. In this context, applications such as videoconferencing, e-mail, enhanced voice messaging, and other information services help to expand the boundaries of the modern enterprise, often without much in the way of a physical presence of that entity in a given market.

Interoperability is concerned with the degree to which we are able to interact with network resources across a range of platforms and architectures. Early efforts by the International Standards Organization fostered the development of the Open System Interconnection Reference Model (OSIRM) (Figure 5-7). The OSIRM is a model built on a layered

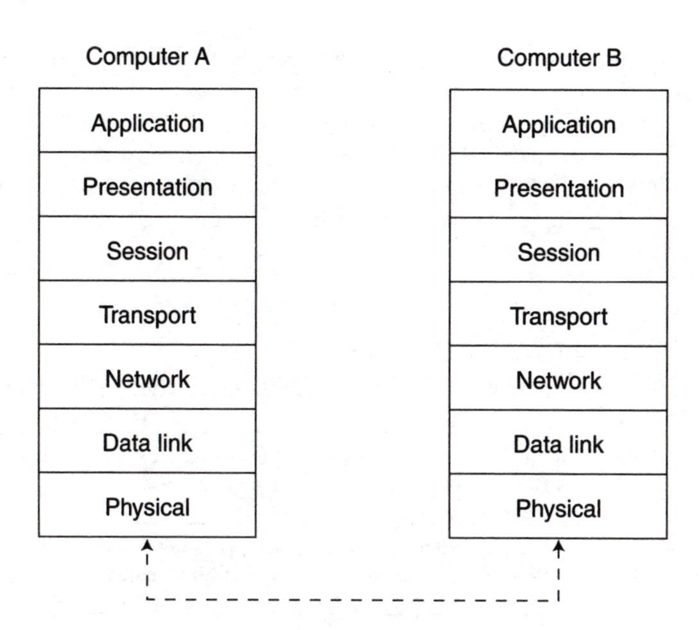

FIGURE 5-7 Open system interconnection reference model

philosophy. In order to understand the significance of adopting a layered approach to network interoperability consider the following analogy.

Recent developments in very large scale integration has flooded the consumer electronics marketplace with a plethora of integrated entertainment devices known popularly as "boom boxes." These boom boxes offer users access to CD changers, tuners, dual cassette tape recorders, and many other elements that were typically found in component audio systems. By integrating these functions into a single device, portability became one of the chief factors contributing to boom box popularity. What was sacrificed by integrating these functions into a single unit was the degree of sophistication each functional unit was able to obtain.

For example, if we were to have a separate tape player, CD changer, tuner, and amplifier, such a system would be more expensive. However, the degree of sophistication associated with each component would be extremely high. By using a component or layered approach to the implementation of a modern audio system, users enjoy the benefits of component specialization and isolation from a single point of failure.

The International Standards Organization (ISO) works very closely with the International Telecommunications Union (ITU) in defining data communications and computer architectural standards. The ISO promulgates standards such as the Open System Interconnection Reference Model (OSIRM). The OSIRM specifies a seven-layer architecture that is fast becoming the de facto basis for global computer-to-computer communications as shown in Table 5-2.

Table 5–2 The Open Systems Interconnection Reference Model

#	Layer	Description
7	Application	The function or process used by the end user of the network. Coordinates security, rules, and conventions between communication entities.
6	Presentation	Reconciles differences in internal machine languages, syntax, and semantics, presenting system users with a unified application interface – i.e., screen, functions.
5	Session	Manages the logical requirements for end-to-end data transmission. (Session management.)
4	Transport	Coordinates the end to end management of data transfer. Service selection, optimum data unit size, and flow control.
3	Network	Provides isolation for the Higher Level Functions (HLFs) performed by layers 4–7 from Lower Level Functions (LLFs) performed by layers 1–2. Network routing, addressing, flow control, and error/recovery services.
2	Data Link	The conversion of bit-oriented data (control and information) into units determined by the protocol in use. Error correction, speed matching, and flow management.
1	Physical	Concerned with the transfer of unformatted or raw data bits, cables, interfaces, power issues, and pin assignments.

COMPARISON OF NETWORK TECHNOLOGIES AND APPLICATIONS

Table 5–3 shows a comparison of network technologies and their applications.

Table 5–3 Network Alternatives at a Glance

Network Transport Technology	Supported Data Rate	Network Classification
100-Base-T	100Mbps	LAN
10Base-T	10 Mbps	LAN
ADSL	16 Kbps – 9 Mbps	MAN and WAN
ATM	>25 Mbps	LAN and WAN
BISDN	>1.544 or 2.049 Mbps	WAN
E Carrier – EC	>2.048 – 565.146 Mbps	WAN
Frame relay	56 Kbps – 1.544 Mbps	LAN, MAN, WAN
Gigabit LANs	>1Gbps	LAN
HDSL	1.544 or 2.048 Mbps	MAN and WAN
SONET	51.84 Mbps – 9953.28 Mbps	WAN
ISDN – Narrowband	56 Kbps – 1920 Kbps	WAN
POTS	<56 Kbps	MAN and WAN
SDSL	160 Kbps – 2.048 Mbps	MAN and WAN
Switched 56	56 Kbps	MAN and WAN
T Carrier – North American	1.544 Mbps – 274.176 Mbps	WAN
VDSL	<51 Mbps based on distance from CO	MAN and WAN

Chapter 6

Telecommunications Services

OVERVIEW

Difficulties associated with obsolescence and incompatibility have made many timid about embarking on new electronic ventures. As soon as we make a network upgrade that will put our companies at the forefront of information technology, another technology, service, or application emerges. Many vendors and telecommunications providers have responded with solutions or application front-end solutions that will ease the speed bumps to the information superhighway.

As wired networks such as WANs, MANs, and LANs become the underlying infrastructure of modern enterprise, people outside the technical community are often less concerned with the intricacies of the network rather than the integration of the services they provide. We have become dependent on various network applications available to end user communities in the form of information or telecommunications services. In this context, the term *information services* is used to describe processes such as data processing, information retrieval, transaction processing, Decision Support System (DSS), voice processing, or any form of information manipulation.

Information services are emerging as one of the fastest growing segments of the information industry. Information service products are available to end user communities via in-house networks or are available from other service providers or what are referred to as third party vendor, service bureaus, or value-added network (VAN) providers. The presence of sophisticated customer premises equipment and enhanced information services represent both opportunity and quandary for deciding the right mix of services and facilities.

The users of in-house telecommunications services often enjoy a higher degree of flexibility in terms of the variety of information resources available, methods for access of services, information retrieval, and subsequent packaging of output. The downside of providing a full complement of in-house telecommunications or information services is reflected in the immense network management burden the organization assumes. Economic constraints often hinder the upgrade of technology and applications. Information services products are typically selected on the basis of financial feasibility rather than technological fit. Additionally, end users with specialized information requirements must often be contented with the generic fare available or be dependent on custom standalone applications.

Securing telecommunications services from outside the end user environment opens a world of sophisticated offerings. In addition to raw digital bandwidth, specialized services such as financial, marketing, scientific, sociological, medical, and other databases are typically available from providers. These provide coordinated information resources and provide the remote user with a single point of entry between associated applications. In this scenario, the provider assumes the network management role that includes ongoing network maintenance and information updates.

There are many global area network applications and technologies for the user to choose from. Most enterprises do not have the luxury of scraping their respective investments in information infrastructure for newer high-speed offerings. In addition to the need to maintain current investments in existing technology, each organization must grapple with the impact of change on the organization and existing customers while balancing the incremental demand to provide requisite technical expertise to end users.

The benefits of global projections (the overall potential market for growth in the range of 30% per year) will only be garnered by those enterprises who make the necessary investments in infrastructure. A number of factors have set the stage for the widespread availability of transmission offerings.

- Standardization in file formats, business operations/procedures, infrastructure, and computing applications.
- Nearly ubiquitous transmission infrastructure – The deployment of analog and digital, wired and wireless connectivity alternatives around the world provides a ready transport mechanism for interactive multimedia traffic.

- Similarities in how enterprises use information technology around the world – The efforts of standards-setting organizations, joint ventures, strategic alliances, and voluntary industry compliance have fostered a high degree of similarity in the ways international organizations use information and associated technology.
- Increased flow of transactions and capital across geopolitical boundaries – Distributed processing has made the international flow of funds and transactions fairly commonplace.
- Information has become the new international currency – A radical shift has occurred from industrial to service in the global economy. As we approach the end of the century, information has become the global currency of choice.

The combined result of these aforementioned factors will allow modern businesses to respond to technologically sophisticated customer requirements in the global arena by

- Providing the company with the ability to compete globally using the most sophisticated transmission facilities regardless of location.
- Reducing costs through the use of high-speed transmission facilities to speed up workflows.
- Responding to competitive changes in the global environment.

The remainder of this chapter is dedicated to an overview of various telecommunications offerings.

PLAIN OLD TELEPHONE SERVICE

The value of starting small should not be overlooked. In this regard, POTS or plain old telephone service provides an excellent on-ramp to the information superhighway. A rich variety of applications, hardware, and expertise, have matured with more than 116 years of telephone service in the United States. POTS is globally available from domestic and international common carriers and other providers (Figure 6-1). The combina-

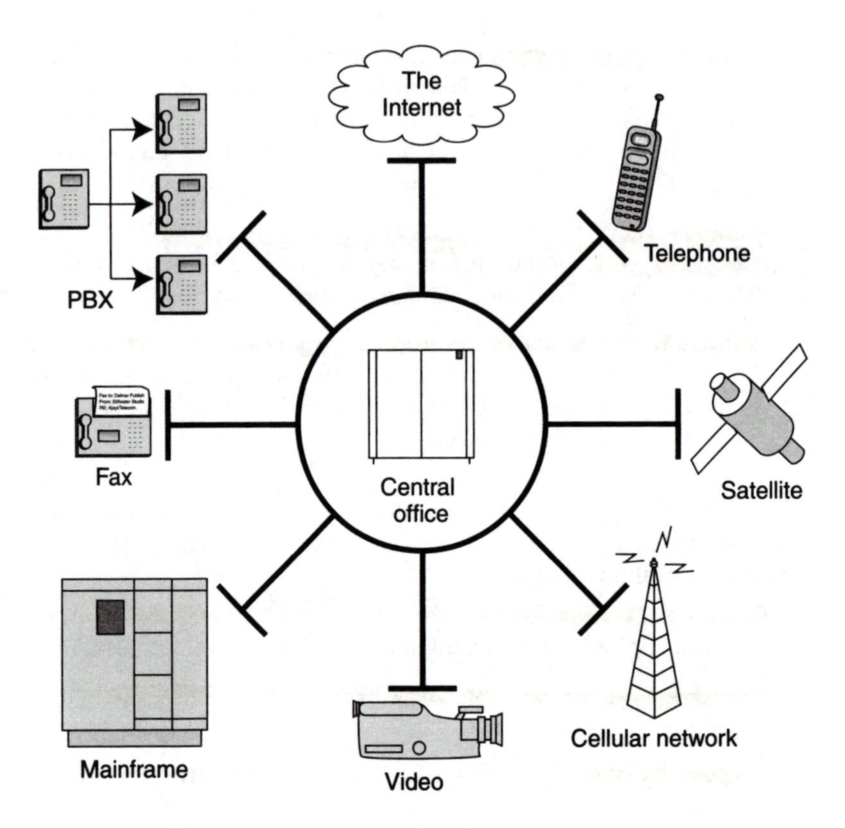

FIGURE 6-1 Plain old telephone system (POTS)

tion of local and long distance service is referred to as the PSTN or public switched telephone network. The PSTN provides narrowband support for voice, data, image, and interactive multimedia applications.

Simple devices such as fax machines, fax modems, modems, and inexpensive line-sharing devices (fax telephone switches) can easily be integrated into existing printing operations with only minor impact to the organization in terms of productivity. The results will be immediate as more transactions are processed with existing resources, faster job completion, better job tracking, and customer database management begin to impact the bottom line.

Many telephone companies (telcos) and more recently, cable TV providers, offer enhancements for both residential and business customers, which allow these groups to add increased functionality and sophistication to existing operations without the associated network management concerns. A sampling of these services include:

- Call forwarding – the ability to toggle with ease between wired, cordless, and cellular telephones around the world. Callers can be switched to voice mail applications or an answering machine rather than encountering a busy signal.
- Call conferencing – the ability to interconnect more than one party during an active telephone session. Usually requires more than one line and can be used in some areas in conjunction with call waiting.
- Call trace – this feature allows law enforcement agencies to track and apprehend obnoxious or obscene callers.
- Caller identification or automatic number identification – caller ID is a service available to residential customers. It allows the called party to determine the identity (name and number) of the calling party. Special services or devices allow calls to be screened and forwarded, and not accepted or disconnected based on subscriber requirements. Automatic number identification or ANI is similar to caller ID. ANI is a service available to business clients that provides comprehensive, customizable data about the calling party.
- Call waiting – subscriber is notified of an inbound call attempt and has the option of placing the active session "on hold" while answering the new call. Recent enhancements allow caller ID to work in conjunction with this feature.
- Call return – allows the subscriber to return the last call received on a given line. Oftentimes, callers will not leave messages in a voice mail system or on an answering machine.
- Voice mail – one of the more recent additions to POTS, which acts as a message center for both business and residential callers. Typically provides increased functionality over answering machines; although in some instances the differences may be negligible.
- Internet services – many local and long distance carriers such as AT&T, Sprint, and MCI offer Internet services. In some instances, industry standard browsers can be used while most common carrier interfaces to the WWW are proprietary.
- Paging – a wireless component of POTS provides for digital and alphanumeric paging. Paging services may be implemented by the common carrier or the customer may manage his or her own paging network complete with base station.

- Cellular – modern cellular communications that offer all of the basic elements of the wired PSTN. Cellular data transmission exists but pales in comparison to its wired counterpart.

Trends in collaborative work, telecommuting, and the generalized availability of affordable switched analog service have been stimulated by high-quality, medium-priced information technology. The typical home (in many cases mobile) office contains a fax, one or more printers, one or more multimedia workstations (networked into a simple peer-to-peer LAN), dual-line answering machine or voice mail, 24X speed CD-ROM, pager, and cellular telephone.

The modern telephone features headsets, caller ID, conferencing, speed dialing, extensive autodial and memory enhancements, wired and wireless.

The benefits of POTS are:

- Ubiquitous availability
- Switched and dedicated backbone for voice, data, image, and interactive multimedia
- Support of both narrowband (<28,800 bps) and broadband (>56,000 bps) speeds

The limitations of POTS are:

- Primarily point to point
- Can be expensive as traffic and hours of usage increase

These noteworthy enhancements in information technology, when coupled with broadband digital transmission facilities, help to overcome the limitations associated with POTS in the printing organization. Digital transmission facilities are available on a demand or switched basis as well as dedicated.

SWITCHED 56

Switched 56 was one of the earliest forms of switched digital transmission. Switched 56 remains a viable alternative for providing connectivity between geographically dispersed customer facilities. As mentioned in previous sections, the term *switched* denotes a connection that is established, maintained, and terminated on a demand basis. 56,000 bps of user bandwidth is provided either by long distance or local common carriers. Data service units (DSUs)/channel service units (CSUs) are required at the customer and destination locations to provide line interconnection to switched 56 services.

T1

T1 is a dedicated broadband digital transmission service provided by various common carriers. T1 service was first offered as a commercial service by AT&T in 1983. Using digital transmission facilities and TDM, T carrier provides digital data rates from 1.544 Mbps to 274.176 Mbps. T1 multiplexers are used at both ends of the T1 connection. Portions of the T1 bandwidth can be partitioned and used on a point-to-point basis known as fractional T1. Bandwidth in excess of T1 speeds can be acquired by using the following T carrier bundles.

North American TDM Digital Offerings

Designation	Number of Voice Channels	Data Rate Supported	Circuit Equivalencies
DSO	1	56,000 or 64,000 bps	—
DS-1 or T1	24	1.54 Mbps	—
DS-1c or T1c	48	3.152 Mbps	2 - T1s
DS-2 or T2	96	6.312 Mbps	4 - T1s, 2 - T1cs
DS-3 or T3	672	44.736 Mbps	28 - T1s, 14 - T1cs, 7 - T2s
DS-4 or T4	4032	274.176 Mbps	168 - T1s, 84 - T1cs

ITU-R – TDM Digital Offerings

Designation Level number	Number of Voice Channels	Data Rate Supported (Mbps)
E1	30	2.048
E2	120	8.668
E3	480	34.368
E4	1920	139.264
E5	7680	565.146

Pulse code modulation (PCM) and TDM are used to transport analog data signals over digital facilities. The input analog signal (4 KHz) is sampled at regular intervals at a rate that is twice that of the highest frequency (8,000 times per second with each sample unit being 8 bits in length). The PCM process produces a data rate of 64,000 bits per second. Using TDM, a T1 line with a rated data speed of 1.54 Mbps can be divided into 24 voice grade or 23 data channels capable of transmitting 64,000 bps.

The PCM process involves four steps:

1. Analog data is input to a codec or digitizer.
2. The output of step one is a pulse amplitude modulation (PAM). T1 multiplexers are used to combine both analog and digital input signals for transmission over digital facilities. One of the major benefits of digital transmission includes its inherent ability to provide a common platform for the integration of various types of data.
3. The PAM sample is quantized or assigned a binary code.
4. The output of step three is an encoded PCM data stream.

SONET

Synchronous Optical Network (SONET) is a fiber-optic-based digital transmission offering providing transmission bundles ranging from 51.84 Mbps to 9953.28 Mbps (Table 6-1). SONET represents an enhancement over other forms of high-speed digital transmission. SONET is based on an international standard called *Synchronous Digital Hierarchy* (SDH), providing users with a standardized digital interface and intelligent net-

Table 6–1 Synchronous Optical Network (SONET)

Service	Designation	Data Rate (Mbps)	Circuit Equivalencies
STS-1	OC-1	51.84	28 T1s or 1 T3
STS-3	OC-3	155.52	3 STS-1s
STS-3C	OC-3C	155.52C	3 STS-1s
STS-12	OC-12	622.08	12 STS-1s, 4 STS-3s
STS-12C	OC-12C	622.08C	12 STS-1s, 4 STS-3Cs
STS-48	OC-48	2488.32	48 STS-1s, 16 STS-1s
STS-192	OC-192	9953.28	192 STS-1s, 64 STS-3s

working support. Many predict that the widespread SONET implementation will likely render T carrier obsolete. SONET is designed to support interfaces to various levels of T carrier, providing a good migration path from those customers who plan to migrate to SONET.

While various transmission technologies vie for dominance in the international telecommunications arena, SONET provides upward compatibility to ATM. ATM uses cells to transport voice, data, image, video, and other forms of interactive multimedia traffic.

ATM

Asynchronous Transfer Mode or ATM is a high-speed digital network service defined as part of the emerging Broadband ISDN (BISDN) standard (Figure 6-2A). As an emerging digital transmission technology, ATM

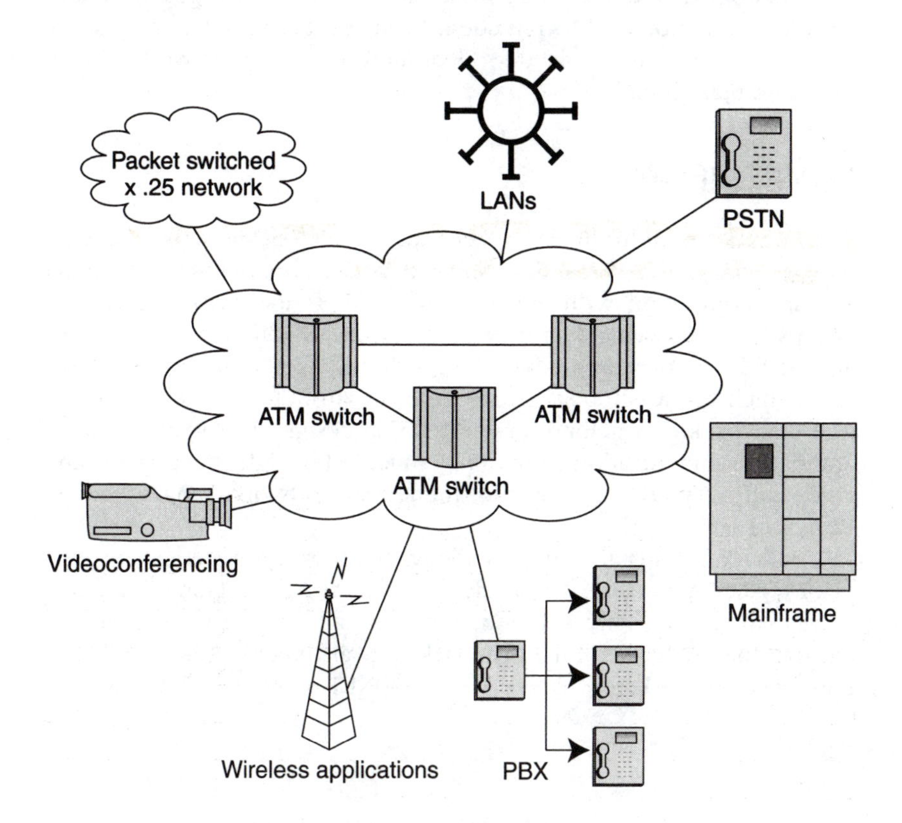

FIGURE 6-2A ATM network

uses fixed cells that are 53 bytes long. The first 5 octets or bytes are used for the header with the remaining 48 used to transport user data (Figure 6-2B). ATM represents the strengths of its predecessors' packet and circuit switching while supporting data rates greater than 25 Mbps. Packet switching was suitable for medium- to high-speed data communications traffic, but proved unsatisfactory for real-time voice and video applications. Circuit switching, while optimized for voice traffic, proved unsuitable for real-time, broadband video, and data traffic. ATM provides a means for universal interactive multimedia traffic between communicating locations using voice, data, and video.

ATM, combining the benefits of switching and multiplexing, is also referred to as a *cell relay technology*. Data—voice, text, video, or image traffic—is broken into cells that are transported across the transmission infrastructure (Figure 6-2C). ATM technology is designed to work optimally in both LAN and WAN environments. Perhaps the largest benefit of ATM lies in its potential to provide a focal point for service integration. This would eliminate the need for dedicated specialized network deployment and management.

FRAME RELAY

Frame relay is a more direct descendant of X.25 packet switching technology and provides for data-only support. As a high-speed digital offering competing with SONET and ATM, frame relay is likely to occupy a niche where data rates close to the established T1 data rates are suitable. Frame relay is based on a "fast packet switching" architecture, which has less transmission overhead attributable to error correction than packet switching. Frame relay is designed to support image and data communications, but is not suitable for high-speed video and voice traffic. Typical data rates for frame relay networks ranges from 56 Kbps to 1.544 Mbps.

As a data transmission technology, frame relay is thought of as a plant replacement alternative to dedicated leased lines. Frame relay networks can be a costly solution, but generally are more robust and cheaper than dedicated lines. The lack of availability will likely relegate frame relay to pockets in larger ATM or SONET networks (Figure 6-3).

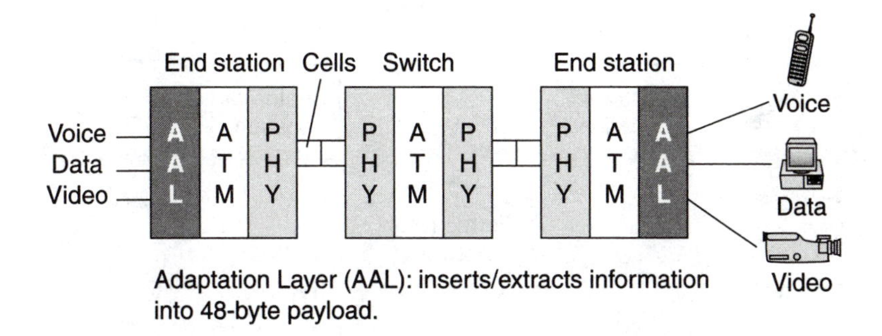

Adaptation Layer (AAL): inserts/extracts information into 48-byte payload.

ATM Layer: adds/removes 5-byte header to payload.

Physical Layer: converts to appropriate electrical or optical format.

FIGURE 6-2B ATM system architecture

Bit

8	7	6	5	4	3	2	1	
GFC				VPI				1 octet
VPI				VCI				2
VCI								3
VCI				PT			CLP	4
HEC								5
Cell payload (48 octets)								6 … 53

GFC: Generic flow control
VPI: Virtual path identifier
VCI: Virtual channel identifier
PT: Payload type
CLP: Cell loss priority
HEC: Header error control

FIGURE 6-2C ATM cell structure

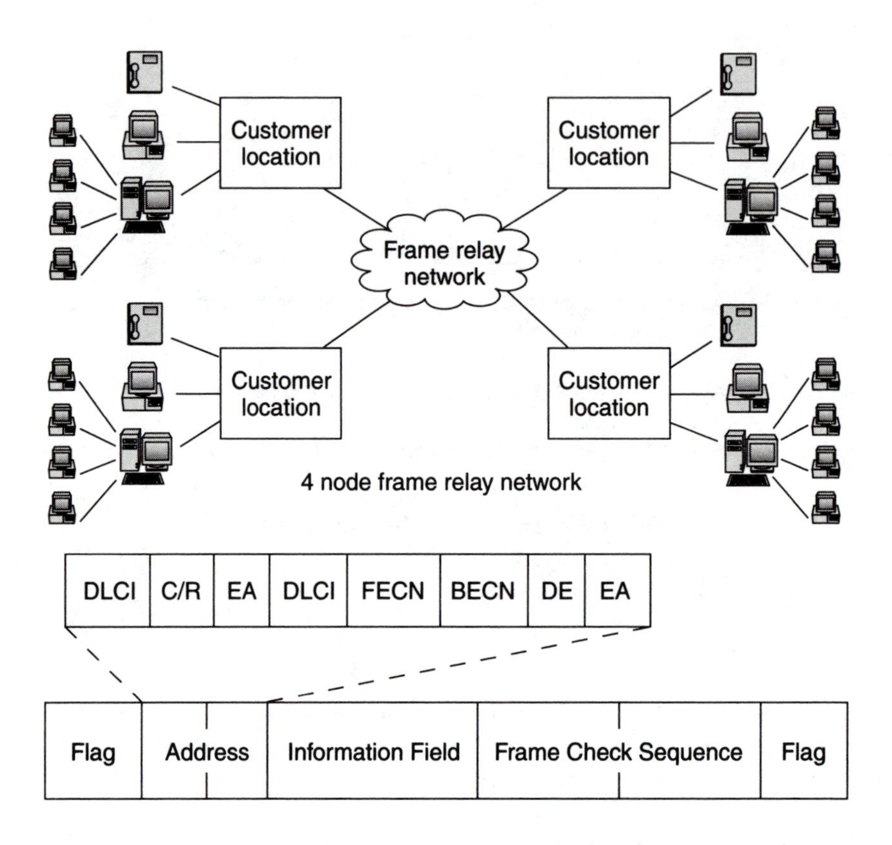

FIGURE 6-3 Frame relay architecture

ISDN

For a long time in the United States, the acronym ISDN was referred to as "it still does nothing." In fact, the printing industry has been one of the largest catalysts for integrated Services Digital Network (ISDN) deployment in the United States. ISDN provides for simultaneous high-speed digital transmission of voice, data, image, and video traffic. As a wideband transport architecture, ISDN provides data rates across a wide variety of speeds (56,000 – >1920 Kbps). The following types of ISDN transmission bundles are used to provide connectivity (Figures 6-4A and 6-4B):

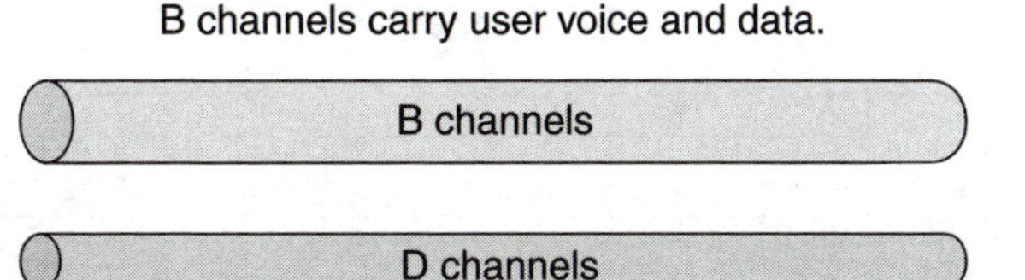

FIGURE 6-4A ISDN channel structures

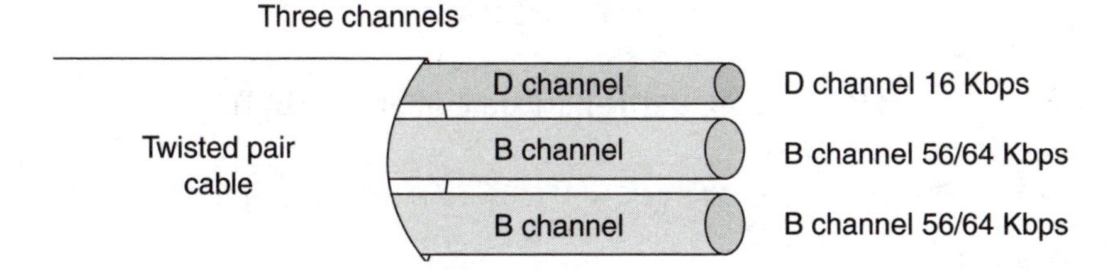

FIGURE 6-4B ISDN B and D channels

- B channels or bearer channels capable of transporting both circuit switched or packet switched traffic at 56,000/64,000 bps
- D channels or delta channels, which transport ISDN signalling or low-speed data (16/64 Kbps)
- H channels, which provide support for broadband data rates (384 Kbps, 1536 Kbps, or 1920 Kbps)

By bundling these transmission channels together, two types of ISDN connectivity are available:

1. Basic rate interface (BRI): The ISDN BRI supports data transport up to 192,000 bps by combining two B channels with one D channel (Figure 6-4C).
2. Primary Rate Interface (PRI): The ISDN PRI supports data transfer rates starting at either 1.544 Mbps (twenty-three B and one D channels) in the United States and 2.048 Mbps (thirty B and one D

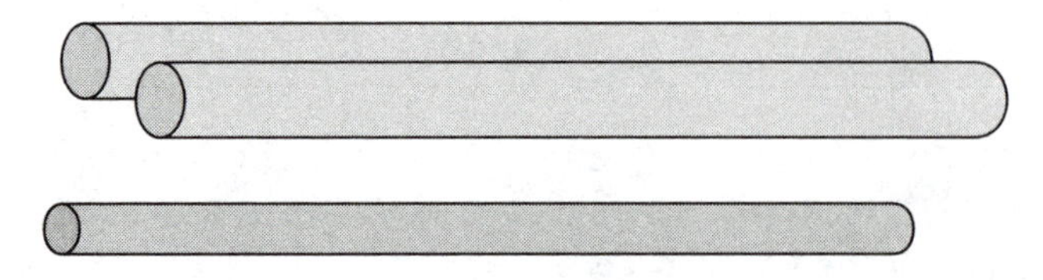

16 Kbps D channel

B channels – user voice, data, image, sound
D channels – call signalling, set-up, user packet data

One BRI = 2 B + D

FIGURE 6-4C ISDN-Basic rate interface (BRI)

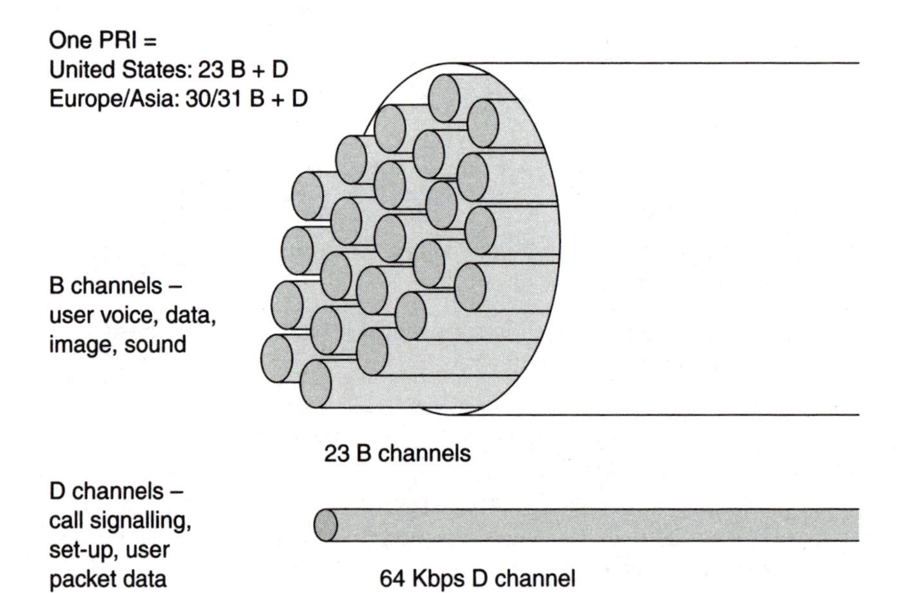

FIGURE 6-4D ISDN-Primary rate interface (PRI)

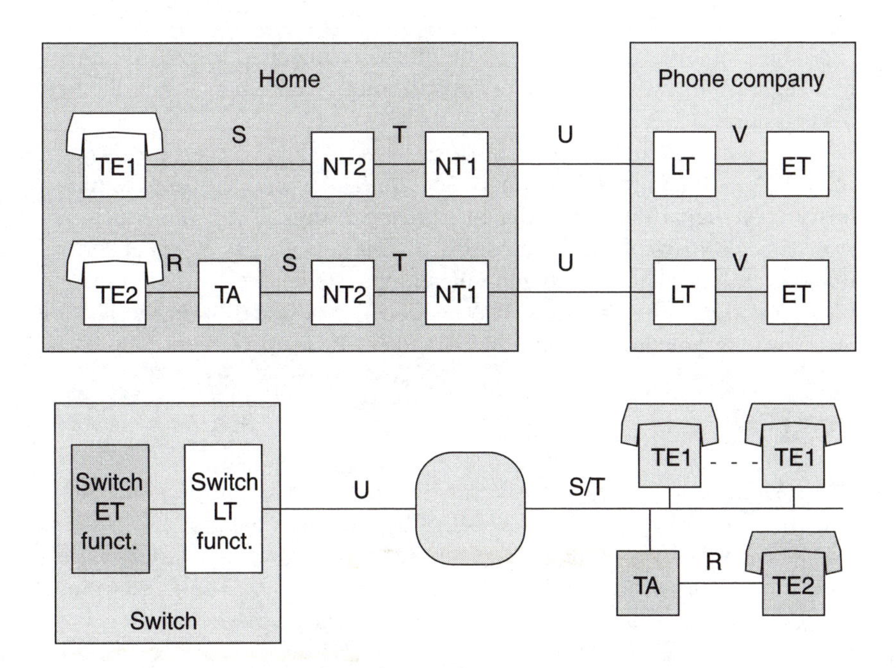

FIGURE 6-4E ISDN network interfaces diagram

channels) in other international ISDN settings (Figure 6-4D). It is important to note that the two iterations of ISDN's PRI are based on differences in bandwidth allocations made using TDM; refer to the TDM discussion in Chapter 2.

For data rates in excess of 1.544 or 2.048 Mbps, subscribers can use H channels for primary rate interfaces to ISDNs. H channels support the following channel configurations

- H0 – Multiple H0 channels are used to provide data rates of 1.544 Mbps using three H0 channels and one D channel or 2.068 Mbps connectivity using five H0 and one D channel.
- H11- A single digital channel with a rated data capacity of 1536 Kbps.
- H12- One 1920 Kbps channel is combined with one D channel.

These channel structures can be mixed or matched to provide error-free digital transmission for specific customer requirements (See Figures 6-4a through 6-4e).

The widespread deployment of ISDN is predicated on the existence of an IDN or integrated digital network. An IDN would support the transmission of digital data over digital signals via digital transmission infrastructure. From previous discussions in this text, it has been observed that currently available telecommunications facilities are a hybrid mixture of analog and digital facilities. The needed IDN foundation in the United States was fragmented, resulting in islands of ISDN implementation.

XDSL

The current push to maintain investments in analog technology and infrastructure has given rise to a new set of transmission technologies collectively known as *digital subscriber line* (XDSL) (Figure 6–5). XDSL is the term used to describe a class of digital subscriber lines. The need to rapidly deploy digital transmission while minimizing costs associated with network upgrades has breathed new life into existing copper twisted pair used as part of the local portion of the PSTN. Between 1995 and 1997, several major corporations conducted a number of trials aimed at determining the viability of deploying a cost-effective, high-speed digital infrastructure using the local loop.

For example a digitized photo 1 MB in size transmitted over switched 56 will take about 17.9 seconds. The same 1 MB photo will take approximately 15.6 seconds using 64 Kbps line. The typical x-ray image of about 60 MB takes about 11.9 minutes to transmit using switched 56 or 10.4 minutes using 64 Kb services. These two file types illustrate the need to provide high-speed digital transmission. However, the incremental costs of implementing such a network would be cost prohibitive. If the same digitized photo was sent in over existing copper twisted pair at 384 Kbps it will take about 2.6 seconds. If we increase the data rate to 1.54 Mbps it would take 0.7 seconds to send the file. By increasing the transmission rate to 384 Kbps, the x-ray file would take about 1.7 minutes. By increasing the transmission rate to 1.54 Mbps, transmission time will be reduced to a mere 26 seconds.

Many common carriers are experiencing heavy traffic and network bottlenecks due to increased Internet usage. The local loop was never

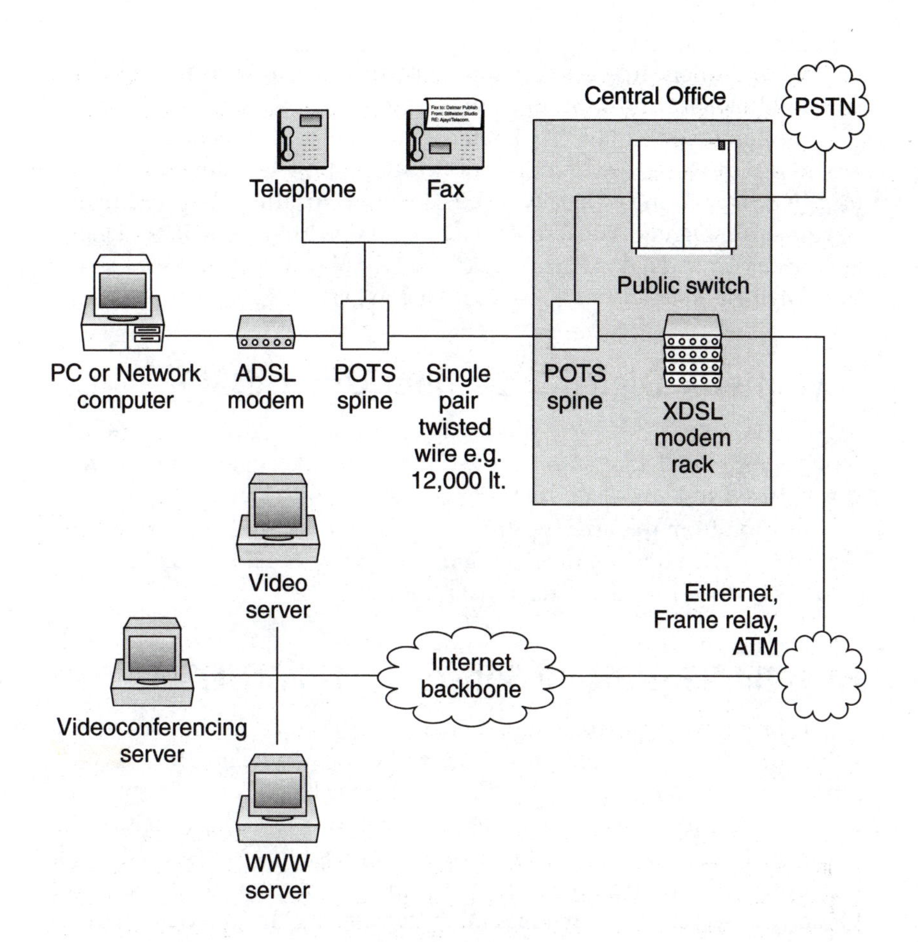

FIGURE 6-5 XDSL

architected to carry the vast array of interactive multimedia applications and data that we access on a daily basis from the WWW. By implementing XDSL technology, many carriers will experience an increase in channel capacity without the cost of major network upgrades. Current activities of the Moving Picture Experts Group is developing a standard called MPEG-4. MPEG-4 specifies data rate of up to 4 Mbps for applications such as telemedicine, telecommuting, video, and videoconferencing. The implementation of MPEG for 1999 along with other bandwidth-hungry interactive multimedia traffic will drive the need for increased digital bandwidth even higher.

As mentioned earlier, XDSL refers to a collection of technologies that would use existing copper local loop to support digital transmission. The four XDSL offerings that will be covered in this section include Asymmetric Digital Subscriber Line (ADSL), High Bit Rate Digital Subscriber Line (HDSL), Symmetric Digital Subscriber line (SDSL), and Very High Bit Rate Digital Subscriber Line (VDSL).

Asymmetric Digital Subscriber Line (ADSL)

ADSL represents a new way of allocating frequency available using subscriber loop. By dividing the bandwidth into three channels, one channel can be used to carry user voice signals. A second channel provides for transmission from the user (upstream) at the rate of 640 Kbps to a receiving entity. The third channel ranging supports variable rates from 16 Kbps to 9 Mbps for downstream data transfer.

High Bit Rate Digital Subscriber Line (HDSL)

HDSL provides for broadband digital transmission. By implementing HDSL, the local loop could provide data rates of 1.54 million bps or 2.048 Mbps. Earlier this chapter we discussed T1, which is a type of digital line based on the North American TDM hierarchy. Broadband transmission in the United States is built on T1 or data transmission groups of 1.54 million bps. In the European community, 2.048 million bps form a basis of broadband digital transmission so this transmission technology provides excellent interconnectivity to, and support for, international digital traffic. HDSL uses existing local loop to provide T1/E1 data rates. Since HDSL is full duplex, users have access to transfer rates between 1.54 or 2.048 million bps in either direction (upstream and downstream).

Symmetric Digital Subscriber Line (SDSL)

Of all of the XDSL offerings, SDSL is perhaps the most versatile. SDSL provides an excellent platform for integrating variable rate digital transmission. Plain old telephone service (POTS) and symmetric bidirectional high-speed traffic can be transported on the same line. SDSL provides variable high-speed transport ranging from 160 Kbps to 2048 Mbps.

Very High Bit Rate Digital Subscriber Line (VDSL)

VDSL provides a range of transmission offerings based on the distance between the subscriber and the central office. Essentially, the closer the subscriber is to the central office the faster the data transfer rate. For example, if the subscriber location is within 1,000 feet of the central office using existing telephone lines, a data rate of 51 Mbps can be achieved. If the subscriber station is 3,000 feet from the central office location, a data rate of 26 Mbps is possible. Finally at 5,000 feet from telco facilities, VDSL can provide an aggregate data rate of 13 Mbps.

Chapter 7

The Internet

INTERNET OVERVIEW

Perhaps no other technology has captured our imagination as the Internet. While the Internet has been in existence for more than thirty years it has only been in the past seven years that intense public and commercial interest has grown around what has been popularly dubbed as the "network of networks" or "the information superhighway." For many, the Internet is a sociological phenomenon very much like the advent of radio, television, and cable TV. The difference between these technologies and the Internet is that the Internet provides a platform for integrating radio, television, cable TV, and myriad other interactive multimedia traffic.

In addition to providing a platform for service integration, the Internet represents a movement toward the popularization of computing. The reduction in the cost of computing technology, the widespread availability of cheap analog and digital connectivity, and the use of graphical user interfaces in the form of Web browsers have made the Internet and its associated WWW the technology of the day.

While many are enamored with and others are skeptical of the Internet, few would argue that if the Internet is not the information superhighway it is at least the on-ramp to something big. The Internet is more accurately defined as a WAN comprised of a complex hybridization of wired and wireless, digital and analog, switched and dedicated, transmission facilities.

As a type of WAN, the Internet is built on the client/server model. In this instance, servers are those computers and other workstations connected to the Internet that provide users or clients with access to application programs, utilities, and other resources contained on them. Clients are a combination of the software and hardware that make up the users environment. Users access the Web using software on their computers or workstations. These clients request information from attached servers, which then deliver requested interactive multimedia content to the clients. Web browsers are currently one of the most popular clients used to access the Internet.

INTERNET DEVELOPMENT

During the 1960s, the perceived threat of global nuclear war caused the federal government, United States military, and the research community to explore highly survivable communications technologies. Meteor burst communications, LAN, and the Internet have common roots in this body of research. Several goals drove the activities of this defense initiative:

1. To provide alternative communications in the event of the destruction of the public telephone network
2. To provide a means of corresponding with essential government, civil, and defense resources
3. To provide a common platform or protocol upon which to build this highly survivable network

During the 1960s, Advanced Research Project Agency Network (ARPANET) was the name given to this initiative. As the threat of nuclear attack became less imminent, the fledgling network established by ARPANET was seen as a means of continuing scientific research by linking academic, government, and other organizations. During the 1980s the National Science Foundation (NSF) extended Internet access to other colleges, universities, and academic resources, connecting them to five large supercomputers. Soon NSFnet extended connectivity to other institutions via seven regional networks.

True to the original research and academic intent of the Internet, businesses and other commercial entities were not allowed interconnection. However, in 1991 amidst the growing strains of managing this vast network and the need to segregate classified information from potential civilian access, NSF separated itself from the Internet. By 1991 the use of GUIs as part of operating systems and application programs made the Internet a new frontier for exploration.

These circumstances allowed companies called *Internet Service Providers* (ISPs) and online service providers to offer commercial Internet services in connectivity to consumer, academic, and business users.

THE WORLD WIDE WEB (WWW)

While many users spend hours surfing the Internet via the WWW, few understand the terminology, technology, and operations of this incredible resource. The following sections provide a basic overview of Internet-related terminology, technology, and applications. Let us start with the basic definition of the Internet. The Internet is the actual network or infrastructure used to connect servers (computers and other workstations) and clients (user computers and workstations). The WWW, also known as W3, provides a means of accessing resources scattered throughout Internet-based resources.

The WWW combines the use of hypertext and graphical user interfaces as a means for navigating the vast information resources scattered on servers and other devices connected to the Internet. Through the use of a protocol called hypertext transfer protocol (HTTP) multimedia-rich Web documents are served to requesting clients via browsers. Using hypertext we are able to navigate this vast maze of information. In essence, when we want to gather more information about a particular topic, we may enter a search or go directly to a previously known location on the Web. We are presented with a list of uniform resource locators (URL) that point to home pages or Web sites that contain the information of interest.

The Web is modeled on several metaphors. The first is a web or spider web similar to the intricate connections made by arachnids. Another way of conceptualizing the Web is to liken it to the pathways or synapses of our brains. As we think, we make connections to pieces of information or data stored in our memory. The Web and the ways in which we search it is seen to follow a similar pattern. Many of us can relate to this because just as we become lost in our thoughts, we often find ourselves lost on the Web as we follow various hyperlinks.

NAVIGATING THE WEB

Behind the mask of graphical user interfaces is a complex series of conventions and addresses that allow us to access e-mail, HTML pages, and other Web-based resources. Many users of the Internet are intimidated by some of these conventions, which include long and often meaningless numbers and names. Upon closer examination, we find that a very simple but ingenious set of protocols has been established to provide transparent access to an increasing number of Web resources. Perhaps the first encounter we had with the Internet was in response to the Web address we sought in the newspaper or TV program. Radio, TV, fan clubs, music, CDs, and interactive participation have prompted us to follow these mysterious pointers. These addresses are referred to as URLs. These URLs point us to Web-based documents that are coded using hypertext markup language (HTML).

There are three parts that make up the typical URL:

1. The first portion of the URL indicates the Internet protocol used to retrieve a document or interact with a server or site. The following is a list of some of the more commonly used protocols that start URLs:

   HTTP – identifies documents that use hypertext conventions or HTML.

   For example: http://www rit.edu/~axaics

   Telnet – this protocol allows users to log onto remote hosts or computers transparently. The user's machine becomes or appears as if it were a locally attached terminal.

   For example: Telnet://www rit. edu

   FTP – indicates anonymous access to the resources on the specified FTP server.

   For example: FTP://www.grace rit.edu

2. The second part of the uniform resource locator or URL specifies the address of the server where the documents we wish to access on the Web are located.

3. The third portion of the URL specifies the exact location of a document or its path in the hierarchy of documents organized on that server or site.

INTERNET ADDRESSING

Now here is the part that makes most of us a little nervous. Remember the first time you tried to configure your Web browser? A polite graphical utility called a *wizard* promised to provide us with an easy-to-follow, step-by-step way of setting up our browser or client. Our excitement over the prospect of sending our first e-mail or accessing tickets via the Web to our favorite sporting event was quickly dampened when the dialogue box prompted us for a domain name server (DNS) address. The charm of embarking on this new excursion rapidly wore off. Many users frightened by the customization process will spend more money for online services that provide transparent access to Web and e-mail. By learning a few basic conventions, users now have a range of services or connectivity options from which to choose.

When configuring a Web browser for the first time, we are asked for a numeric address or what is known as an *IP* address; 127.4.67.193 is an IP (Internet protocol) address. Remember, the Internet is a type of WAN that uses TCP/IP protocol suite as the basis for managing crossed platform interconnectivity. The TCP portion of the protocol suite breaks e-mail messages, home pages, and other requested documents into smaller packets for transmission and receipt between servers and clients on the Internet.

TCP not only breaks our data into smaller packets but it also assigns numbers to these packets and routes them in the most efficient manner to our destination or client. The presence of other users or traffic on the Internet may cause the packets that make up our e-mail message to arrive at our client at different rates and in a different order than they were sent.

The TCP portion of the user or client reassembles these packets into their appropriate order making our e-mail easy to understand. In order for these packets to be requested from the appropriate server and routed to the requesting resource, the IP portion of the protocol suite establishes a number of conventions to manage this process.

The IP address is a numeric pointer to location of resources located on WWW. In addition to IP addresses, domain names are often used that more closely resemble names or identifiers that are similar to those we use everyday. For example, the IP address given previously can easily be represented as:

http://www rit.edu/~ axaics

In this example, HTTP is specified as the protocol, www indicates the server type, rit identifies the hostname, and edu identifies the domain. Other Internet domains include

.edu

.mil

.org

.net

.com

.gov

By combining these conventions, users are able to access home pages, send e-mail, or interact with applications in utilities that reside on a specific server located on the Web. As stated previously, access to servers and the information contained on them is governed by the protocol suite known as *TCP/IP*. TCP/IP is a layered protocol used in conjunction with the specialized functions provided by other TCP/IP and non-TCP/IP protocols.

Internet protocol version 4 (IPv4) and Internet protocol version 6 (IPv6) both specify a 32-bit or 128-bit address length respectively. IPv4 is currently the most widely used version of the Internet protocol and forms the basis of this discussion. There are two basic parts of an IP address. One portion of the IP address identifies a specific host or machine in a given network. Another portion of the IP address identifies a specific device or user within that network. In order to locate resources on the Internet, a unique network number and local number are required. Internet addresses, or IP addresses, are assigned to organizations on the basis of their size. Four basic classes are used in IP addressing:

Class A – assigned to large networks

Class B – assigned to intermediate-sized networks

Class C – assigned to small networks with <256 devices

Class D – used for multicast

Other Protocols Used with TCP/IP

In addition to the functions provided by TCP/IP, other protocols provide e-mail, news, file transfer capabilities, and other essential services requested by clients attached to the Internet. The following section will look at a few of the most commonly used protocols.

Simple Mail Transport Protocol (SMTP)

SMTP is part of the TCP/IP protocol suite that manages the transmission and reception of e-mail messages between users. SMTP ensures that the appropriate messages are routed to the appropriate users. Two other protocols are typically used in conjunction with SMTP. They are post office protocol 3 (POP3) and Internet message access protocol (IMAP). POP3 allows users to read e-mail in their mailboxes or server. Using POP3, a user's e-mail is transferred to the user client and is deleted from the server. IMAP provides the user with more flexible options for reading e-mail. Using IMAP, the user can download his or her e-mail, read it, and decide whether to delete it from the server.

File Transfer Protocol (FTP)

While e-mail is the most popular use of the Internet, many sites or servers contain utilities, programs, and other resources that can be downloaded to user clients. FTP provides a transparent way for software vendors to provide fixes or updates to application packages via the Internet. Documents that may be in other formats such as portable document format (PDF), sound files, video clips, and other interactive multimedia format, may be too large to view or manipulate in real time. By downloading them and saving them on the client workstation, users can run these applications at their convenience.

Telnet

Telnet provides a transparent means for clients to log onto other remote computers using the Internet as a backbone. Essentially, Telnet allows the user to log onto the remote host as if they were locally attached. In order to establish a connection to the remote host the user will require a Telnet client. Many operating systems include a Telnet client or one can be downloaded from a shareware site. Web browsers may also be used to establish Telnet sessions on other remote hosts.

GETTING CONNECTED

Now that we have explored some of the technical nuances of the Internet and WWW, let us focus our attention on the ways in which we gain access to these resources. Internet service providers (ISP) provide access. Online service providers provide access to the Internet and they may also provide content in the form of specialized application programs or utilities. For example, when using America Online (AOL), users are provided with e-mail access, Internet access and specialized databases packaged especially for AOL users.

ISPs provide users with either dial/dial-up or dedicated access to the Internet. Many commercial organizations use both dedicated and dial-up access to the Internet. Inexpensive Internet access is provided to non-commercial users through dial connections in the form of serial line Internet protocol (SLIP) or point-to-point (PPP).

SLIP and PPP are just two types of Internet protocols that support two different levels of connectivity. They differ in terms of the error-detection capabilities, mode of communication supported, and line-sharing capabilities.

Point-to-Point Connection (PPP)

PPP supports multiple-user access to the Internet. PPP can also provide connectivity to both asynchronous and synchronous devices. PPP also provides a high degree of flow control and error detection.

Single Line Internet Protocol (SLIP)

SLIP is another means of gaining dial access to the Internet. SLIP connections are typically slower than those provided by PPP and are essentially no-frills connections to the Internet.

ABOUT HOME PAGES

Using the hypertext markup language, commercial, academic, and other users of the Internet can have a presence on the Web in the form of Web sites or home pages. For many, the creation of a home page appears to be a mystical process. Many shudder at the thought of coding text, images, voice, and other content using HTML. Not only is HTML not difficult to learn but for those of us who are the bit faint of heart we can use application programs, which provide us with "what you see is what you get" (WYSIWYG) Web page creation with drop and drag ease. Most of the current generation of wordprocessing software also provides built-in HTML translation utilities.

Using your word processor you simply create the page as you would like it to look. By clicking or selecting the appropriate item from the toolbar, the page or pages are automatically translated into HTML documents suitable for a home page or Web site. How or where do we place these newly created Web pages or sites? Many users do not realize that in addition to the Web access provided by their ISP, many offer their clients a partition (space on the server) into which they can upload these pages or Web sites. Contact your Internet service provider for details.

INTERNET COMMERCE

One of the key issues and attractions of the Internet has focused around electronic commerce over the Web. Many users feared that the 1991 NSF pullout from the Internet would result in the Internet being overtaken by rank commercialism. Nonetheless, many companies are racing to explore this new venue for their goods and services. Many of the controls that exist or govern retail and other forms of electronic commerce may not apply to the Internet. Also security looms as a potential problem for Internet transactions. The very nature of the standards that make it possible for those attached to the Internet to communicate regardless of the platform also makes it vulnerable to hacking and other illegal activities.

Federal agencies, financial institutions, credit card companies, and other stakeholders in electronic commerce have banded together to establish a set of guidelines, security measures, and penalties for Internet commerce.

The commercialization of the Internet has opened a new frontier for buying, selling, and bartering goods, services, and other electronic wares. Each day an increasing number of vendors are hawking their products on the Internet. The lack of clear-cut standards has left some businesses fearful of participation due to the potential for fraud. Others are fearful that they might lose out if they do not get in on the ground floor. Countless articles, books, and editorials form the daily flood of information on television, radio, newspaper, and other forms of electronic media about Internet (I) commerce.

Simply put Internet commerce is a form of electronic commerce. Electronic commerce refers to the exchange, purchase, sale of, and bartering of goods, information, services, and other electronic wares, using both public and private networks. Electronic commerce has its roots in electronic funds transfer (EFT). Through a complex arrangement of interconnection agreements and standardization, we are able via automatic teller machines to access, manipulate, and withdraw funds from our various business and personal accounts. It would only stand to reason that the Net would become an extension of this system.

Internet commerce has become the fodder of countless reports, industry overviews, and strategic analysis. There is little agreement not only in the definition of Internet commerce but few agree on the types of transactions that are currently carried out on the Web. The following list is a synthesis of varying viewpoints regarding the nature of transactions that are currently being executed on the Web.

- Financial institution to financial institution. Electronic funds transfer (EFT) and electronic data interchange (EDI) have been used for many years between financial institutions, businesses, and suppliers. The same types of transactions are now appearing on the WWW. They include transactions using wire transfers, credit card, and what is known as *digital cash*.

- Hybrid forms of electronic banking. Other forms of electronic banking that have recently made their appearance on the Internet including stock transfers/trading, electronic bill paying, wire transfers, and credit card application just to name a few.

- Business to consumer. This area or type of Internet transaction is the one that is being most closely watched. With an estimated one-third of all homes in the United States having access to some form of computing, the potential for this segment of Internet commerce is staggering.

- Consumer to consumer. Consumer to consumer is perhaps the most overlooked potential for Internet commerce. In this light, small cottage businesses, electronic flea markets, and trades among individual users are becoming more commonplace. In addition to tangible objects being exchanged, services such as astrological readings, dating services, and Web page creation have begun to emerge and has catapulted many into the category of electronic or Internet millionaires.
- Inter-enterprise or business to business. This type of transaction most closely resembles EDI except that it is conducted on the Internet. The difference between EDI and inter-enterprise transactions is that EDI transactions typically occur in a closed loop or network environment with associated internationally recognized standards.
- Intra-enterprise. This type of Internet transaction is more likely to occur in Intranets or private TCP/IP-based networks that use HTTP and other Internet protocols to provide HTML-based documents and services in-house. Connectivity between dispersed locations of an enterprise might use the Internet as a means of conducting intra-enterprise transactions.

So what are some of the issues that users, enterprises, financial institutions, and the government must consider in light of the increased interest in Internet commerce? There are many. The following list summarizes some of the salient issues that have surfaced in the discussion of Internet commerce.

- Anonymity – the need to maintain the confidentiality of all parties involved in a given transaction. They include the seller, buyer, and any financial organization involved in the transfer of funds related to a given transaction.
- Authentication – the ability to determine access and authority to conduct a particular transaction. The goal is to be sure that the parties involved are those who are authorized and legally recognized.

- Availability and reliability – this is a major concern in Internet commerce because increased traffic on the Net has made it less than trustworthy as a venue for high-volume, real-time transaction processing. Availability refers to the period of time that the Internet is accessible whereas reliability is concerned with the integrity of the transaction.
- Collection of funds – Internet "funny money" or digital cash are among the many ways that customers and enterprises can pay for the goods and services exchanged. While credit card transactions make up the bulk of Internet commerce new technologies and techniques are being explored to safeguard against theft and other forms of electronic fraud. Another issue that remains problematic are the lack of clear-cut remedies that are available to both consumers and enterprises in terms of refunds or collection of funds for items or services received.
- Data integrity – is concerned with and is closely related to issues of authentication and security. The goal is to ensure that all Internet-based transactions are conducted over secure facilities, that the data contained in the transactions is unaltered, and funds for such transactions are transferred in a timely and accurate manner.
- Security for Internet transactions – encompasses anonymity, authentication, availability, reliability, collection of funds, data integrity, and transaction validity. The ability to ensure, protect, or enforce penalties in an open environment such as the Internet remains problematic.
- Transaction validation – is concerned with ensuring that all Internet transactions occur between valid trading partners, using electronic or other currency, and that the consumer receives what was purchased and the supplier receives remuneration for items purchased.

The following are some security techniques for Web-based or Internet commerce:

- Encryption – encryption involves rendering Internet transactions unreadable or otherwise inaccessible to unauthorized parties. In the area of Internet commerce, algorithms such as Rivest, Shamir,

Adleman, encryption (RSA), pretty good privacy (PGP) and other techniques are used. A combination of public and private keys are used to encrypt and decrypt Internet transactions. Remember, keys are specialized algorithms used to encrypt and decrypt data.

- Digital signatures – in the real world authentication for retail, catalog, or other transactions is provided by an individual signature. This same concept has made its way into electronic commerce as a means of up authenticating, ensuring integrity, and confidentiality of transactions. Essentially, digital signatures are unique random numbers that are created through a process called *hashing*. This uniquely created number is then encrypted using a private key and is attached to the original transaction.

- Digital certificates – digital certificates are another type of popular security method used to protect documents and other transactions on the WWW. Essentially, a certificate is a password-protected, encrypted file that contains information related to both public and private keys, identification of the user or valid users, the name of the certification authority, and the period for which the digital certificate is valid. A certification authority acts as a third party to ensure that the public key is authentic.

- SSL – stands for secure sockets layer. Using RSA public key encryption, inter-layer communications between protocols and client and servers is protected. Data and other control information are exchanged between the aforementioned entities in what is known as *sockets*. Essentially, data integrity and security are built into the program layer of the application and the Internet. For example, Netscape uses SSL as a means of managing the security and confidentiality of data transmitted and received by this popular Web browser. For example, users are prompted when they are about to send or receive data from an unsecured site.

- SET – secured electronic transactions is based on SSL protocols for providing security in the Internet environment. Through the collaborative activities of companies such as MasterCard and Visa, SET has emerged as the de facto standard for providing security for Internet-based transactions. SET uses digital certificates that are issued by banks and other financial institutions. Remember, the certificate is the file that contains information used to authenticate the identity of the user.

Warranties, Contracts and Liability

The newness of Internet commerce has created many new opportunities as well as challenges. Perhaps one of the chief challenges lies in finding a way to provide for the same legal protections for transactions conducted on the Internet that are found in other warranties, contracts, and other legal documents. The legal community found their hands full when ATM, EFT, and 800 services were used for modern commerce. Some of the issues that plague the ability of lawmakers and the courts in protecting the rights, guarantees, and warranties, and responsibilities of consumers and enterprises are

- Lack of clear-cut jurisdiction for enforcement
- Lack of technical expertise in the courts and other constituents of the legal community
- Lack of precedence in this new electronic frontier

While financial institutions, enterprises, various levels of government, and the legal community have banded together to explore the challenges and issues related to Internet commerce, many are concerned that the rights of the consumer will go by the wayside. Issues such as privacy, unwanted solicitation, and taxation are now being applied to Internet commerce. Recently, governors in the United States have taken exception to the Clinton administration's desire that the Internet remain a tax-free zone. They have proposed that a uniform tax be applied to all transactions conducted on the Web.

Chapter 8

IT at Work

IT IN THE WORKPLACE

Information technology (IT) has become the fabric of modern business. Applications such as voice mail, e-mail, and network computing have extended the boundaries of commerce beyond physical walls. Telecommuting or what is now known as the *virtual workplace* has become a reality. Many individuals complete complex work-related tasks in the comfort of their homes. The typical home office might include a Macintosh or IBM computer, laser or color printer, fax machine, copier, voice mail, and other office technologies.

The widespread availability of high-quality analog and digital facilities makes it possible to connect sophisticated telecommunications services to geographically dispersed workers. Mobility and portability have been major catalysts in the creation of the mobile office. Typical mobile configurations include most of the functionality found in a corporate office. However, the integration of wireless technology with robust notebook computers have made "office on the go" a reality.

The following sections will provide a snapshot of some of the tools and applications that have made computer-supported collaborative work (CSCW) and groupware possible. Computer-supported collaborative work has emerged as one of the new paradigms for the modern workplace. Technologies such as local area networks, software office suites, e-mail, and other forms of messaging allow coworkers to concurrently access and manipulate shared information resources. *Groupware* is a name that is given to the collaborative tools that foster CSCW.

E-MAIL

E-mail or electronic mail is perhaps one of the most popular applications on the Internet and online services. E-mail is a type of asynchronous messaging (where asynchronous means the communicating parties are not present at the same time). The simple mail transport protocol (SMTP) layer of the transmission control protocol/Internet protocol (TCP/IP) protocol suite governs the rules that manage the transmission and receipt of electronic mail.

As a type of store and forward technology, e-mail uses SMTP as a means of transmitting and storing queued messages on an Internet or Intranet server. Using an e-mail client such as Eudora Pro or the mail feature of Netscape, users can retrieve messages that are being held for them by the server. Interactive mail access protocol (IMAP) allows users to view mail on the remote server and decide whether or not to leave the mail on the server even though it may be deleted locally. For example, rhianon@frontiernet.net is an e-mail address for the user rhianon whose mail is kept on the frontier server. The .net extension identifies the server/domain as a network provider.

Other useful domain name extensions include:

.edu	educational institution
.com	commercial entity
.mil	military
.gov	government
.net	network provider
.int	international organization
.at	Austrian
.au	Australia
.ca	Canada
.arts	art and culture
.rec	recreation and entertainment
.nom	individuals
.info	information services

While the majority of e-mail messages contain simple ASCII text, e-mail client plug-ins support the inclusion or attachment of audio, video, sound, and other interactive multimedia files. With these enhanced capabilities, a new degree of interactivity has been added to this simple form of communications. While e-mail has become an invaluable part of business communications, many users are beginning to recognize some of the limitations associated with it. Even with the attachment of interactive multimedia files, e-mail still lacks the context or meaning that might be conveyed over the telephone or in face-to-face meetings.

Smileys are used to temper the tone of an e-mail message. A few smileys are included for your review:

:-)	smile
: ->	sarcasm
: -)	laughing tears
; -)	wink
: -<	sad

VOICE MAIL

Voice mail is another form of asynchronous messaging. Voice mail can be implemented using a card in the computer, a standalone voice mail system, or purchased as a service from a common carrier. Voice mail, along with e-mail, are perhaps the two most frequently used resources in the virtual workplace. While oftentimes confused with an answering machine, voice mail is an extension or type of automated attendant with a more personal touch. Usually a caller is greeted by the voice of the person whose mailbox was reached.

In addition to recording voice messages, most voice mail systems allow users to bypass the system in order to interact with a person in real time. The ability to route to multiple voice mail messages at the same time increases productivity. For example, to schedule a one-hour meeting with five colleagues, after a number of telephone calls, e-mail messages, and attempts to physically locate them, more time has been spent coordinating the meeting than the actual duration of the proposed meeting itself.

VIDEOCONFERENCING

Videoconferencing is perhaps one of the most interactive but costly means of providing visual connectivity among people who are geographically dispersed. Using transmission facilities such as T1, ISDN, switched 56, or other digital offerings, multiple locations may simultaneously receive full motion, color, CD quality audio, and images. The use of video compression techniques allow for lower resolution videoconferencing over narrowband facilities. By using videoconferencing technology, we are able to conduct meetings with individuals at remote locations while maintaining a sense of communicating "person to person."

While the cost associated with videoconferencing has dropped dramatically, the provision of full-motion, full-color videoconferencing is out of reach for many organizations. Hence, desktop videoconferencing using the Internet or LANs as a backbone has garnered a great deal of interest. At the desktop level, users can interact visually. Each user must have a small video camera, access to the Web, and a videoconferencing product such as CU-SeeMe. Used alone, the current generation of desktop videoconferencing over the Internet or LANs is not sufficient to meet the rigorous requirements of modern business. Used in conjunction with other collaborative tools such as e-mail, voice mail, fax, and teleconferencing, desktop videoconferencing occupies a growing niche in the business world.

ON-DEMAND PRINTING

As we approach the turn of the century, the printing industry will continue to explore the challenges and opportunities of integrating electronic communications into the printing workflow. Perhaps the need for "just in time" or distributed on-demand printing will continue to drive this revolution. There has been a generalized trend or shift from the print-then-distribute model to one that embodies timeliness, flexibility, and high quality. These factors previously made the production of less than 5,000 of a given piece cost prohibitive for most medium-to-large print shops. Meanwhile, developments in mass storage, high-speed file transmission using inexpensive, high-quality modems, and competitively priced digital transmission now proffer a technological solution that will make short run distributed printing a more attractive proposition.

In fact, between 40% and 50% of all commercial in-plant printing demand is characterized as short run black and white or color-printed pieces. This demand boasts revenues between $36 and $45 billion in the

United States alone. In order to fully appreciate the likely impact of electronic communications on the printing industry in the decade to come, we must examine the ways in which it has already transformed the industry.

In 1994, more than 36 million personal computers were installed in households in the United States with estimates of more than 57 million by the year 2000—or growth of more than 60% annually. During this same period, it is reported that there are more than 19 million modems in use with an annual growth rate of 165% or 50 million modems by the year 2000. When we consider the penetration rates of telephone and cable television in the United States, it becomes clear that more than 70% of all households have access to an ever-increasing information universe via a variety of telecommunications facilities.

The impact of electronic communications on the printing industry and its associated processes can best be examined in the following three phases:

- Phase I: Electronic communications in the form of desktop publishing and wide area and local area networking allowed many print shops to streamline operations. Local computers and workstations were interconnected to high-quality laser printers (color and black and white), scanners, imagesetters, and other devices, resulting in increased workflow. People with specialties or who performed specific tasks could be shared on a local or global basis. For example, many of the magazines, newspapers, or other periodicals that we read were created by freelance writers, photographers, graphic artists, and production staff from around the world.

- Phase II: During this phase electronic communications in the form of high-speed, dedicated and switched digital transmission facilities (T1, T3, switched 56 or ISDN) were used to enhance distribution of finished materials to print shops and plants around the world. Many large publishers reaped enormous benefits from transmitting completed jobs to production houses around the world, offsetting cost increases associated with escalating ink, paper, and postage prices. The producers of these variable print publications were able to create essentially custom materials based on demographic, regional, or other customer-defined elements.

- Phase III: In this phase, electronic communications will likely have the biggest impact on the printing industry in the areas of remote proofing, computer to plate (CTP) and the transmission of customer files for final production. In addition to traditional offset, modern digital print shops will likely derive a significant portion of their revenues from the variable short run distributed on-demand printing.

In order to fully integrate these services into the current offerings, printers must sift through a seemingly impenetrable maze of connectivity options and transmission alternatives such as ISDN, ATM, SONET, frame relay, and switched 56 services.

The world of telecommunications and computers, collectively referred to as *information technology,* is not for the faint of heart. Many printers are dismayed by the endless melange of acronyms and technical details typically associated with just learning about the differences between applications and technology.

Next, many would-be participants find themselves bombarded by an endless array of vendors and common carriers hawking "end-to-end" solutions, which end up being proprietary technological nightmares for them and their customers. Many production houses are wary of future investments in infrastructure to support the volatile distributed on-demand printing market development after rather lackluster productivity gains with information technology (IT) investments made in the 1980s and 1990s.

Difficulties associated with obsolescence and incompatibility have made many printers timid about embarking on new electronic printing ventures. Many vendors and telecommunications providers have responded with solutions or application front ends that will lessen the speed bumps to the information superhighway. There are many global area technologies and solutions available to the printer to facilitate their navigation through the aforementioned phases of integration.

So who stands to benefit from the integration of information technology into the printing workflow? On-demand distributing printing, CTP, and other forms of electronic file transfer will prove to be a boon for:

- Customers – flexibility in producing printed materials for internal and external information requirements will function as an enabler to printing customers in their respective enterprises.

- Printers, advertising agencies, and other graphic arts organizations – benefit from market expansion through an electronic presence on the information superhighway, the Internet, and related electronic venues. The ability to share information and expertise with others in varying stages of integration and the creation of new customer services available on both an in-house and distributed basis are additional benefits.
- Telecommunications/information vendors, manufacturers, and service providers – collaborative development of products, infrastructure, and services that foster cross-platform interconnectivity and interoperability for printers and their customers.

IT IN MEDICINE

Information technology has made major contributions to the study and practice of modern medicine. A visitor to a modern intensive care unit will find critical care and monitoring provided for patients through the use of state-of-the-art information technology. Information about a patient's status, which include arterial blood gas, pulse, rate of respiration, and other vital statistics are gathered and relayed using wireless technology to monitors in the doctor's/nurse's station. When some predetermined threshold is exceeded or is too low, audible and visual alarms alert medical personnel of the status of multiple patients.

Information technology not only plays a major role in monitoring but associated disciplines such as artificial intelligence, robotics, and expert systems assist health care providers by providing automated support in the form of pharmaceutical dispensing and diagnosis. What is it about medicine that makes it an excellent candidate for the widespread deployment of information technology? The following factors have been identified as major catalysts in the integration of information technology in health care:

- A generalized trend toward a shrinking base of geographically dispersed medical expertise.
- Modern health care and the critical nature of medicine demands rapid response characterized by a high degree of accuracy.

- Mobility and portability have become necessities in providing health care.
- The need to control escalating medical costs have made the need to provide efficient health care through consolidation, strategic alliances, and reduced infrastructure, a major balancing act.

The technologies that continue to play a major role in medical education and health care include the following:

- High-speed computing. Advances in integrated circuits have resulted in the miniaturization of computing devices. Consequently, today's computers are smaller, faster, and more robust. More importantly, reduction in the costs associated with IT and medical applications has made computer technology nearly ubiquitous in the health care industry.
- Imaging. Imaging is perhaps the most significant development IT has contributed to medicine.
- Network computing. The need to connect geographically dispersed information resources, medical professionals, and the wealth of medical knowledge derived from centuries of medical expertise, has been a major movement in the provision of health care. Through the use of networked medical resources, practitioners have access to instantaneous information from around the world.

Key IT Medical Applications

The following section outlines some of the key medical applications supported by modern information technology.

Remote Diagnosis

Network access to large medical databases and real-time access to medical experts around the world make it possible to provide consultation and diagnosis for a patient based on transmitted information. The idea of performing medical examinations and evaluations using telecommunications networks is not a new one. Shortly after the invention of the telephone, attempts were made to transmit heart and lung sounds to experts who could assess the state of the organs and the patient.

Teleradiology

A good portion of modern medicine is dependent on imaging techniques and their tangible byproducts. Hard copy still remains the most common form for distributing medical images for diagnostic purposes. The ability to transmit x-ray, MRI, and other visual information in digital format has been facilitated by the use of T1, ISDN, and other high-speed digital transmission facilities. Essentially, the original or hard copy versions of x-rays and other medical images can be digitized using specialized scanning technology. Once digitized this information can be transmitted to remote sites instantaneously.

Telepresence

The combination of virtual reality, expert systems, and a greater understanding of human computer interaction have made telepresence a reality. Telepresence is described as a system and associated applications that allow medical practitioners to share expertise, problem-solving skills, and sensory motor facilities with others at remote locations. Essentially, a doctor is able to perform some procedure or task at a remote location while having the sense of actually being there. This is achieved by providing sensory motor cues and feedback to the physician. The doctor is said to be virtually present in the remote location.

Automated Pharmaceutical Dispensing

A number of technologies expand medical intervention by dispensing medication based on predetermined criteria. Automated pharmaceutical dispensing (APD) assists medical practitioners in the calibration, preparation, measurement, and distribution of prescription medicine. APD provides for inventory control, prevents the contamination of sterile utensils, minimizes the risk to humans who work in hazardous biomedical environments, as well as providing increased efficiency in dispensing medications with greater consistency in doses.

Augmented Reality

The Internet or any attached medical Intranets are being used extensively to provide medical education and training. Augmented reality allows training institutions to be connected to hospitals and other active medical environments. One of the benefits of augmented reality is that medical practitioners are able to conduct real-time interactive consultation—what is known as an *electronic second opinion*. Augmented reality supports the real-time visualization of many precision surgical techniques. Students have the ability to enhance their skills in anatomy and physiology using augmented reality.

Telemedicine

A number of processes have been identified under the heading of telemedicine. They include the transfer and manipulation of radiological and histological data, remote consultation, remote diagnosis, remote decision support, and diagnostic expertise.

Computer Assisted Surgery (CAS)

The widespread availability of high-speed digital transmission facilities, advances in imaging techniques, and network computing have introduced a number of new tools into the surgical environment. Interactive real-time data in the form of x-rays, microscopy, and MRI scans can be transmitted in real-time to surgical experts and practitioners around the world.

DISTANCE LEARNING

The need to provide quality education to an increasingly diverse range of students has made distance learning an invaluable tool in preparing professionals for the Information Age. Over the years, distance learning has lost its stigma. Major changes in commerce, vertical integration, and the need to provide highly trained professionals across a range of industries, prompted a major philosophical shift in the way higher education is provided.

While the traditional on-campus environment continues to be the primary model for providing university or college education, the virtual campus (made possible by distance learning) is becoming an accepted and strategic part of the academic landscape. The notion of distance learning is not a new one. Radio communications provided one of the earliest forms of distance learning in Australia. By combining synchronous and asynchronous technology and applications such as e-mail, voice mail, enhance messaging, videoconferencing, and a host of other interactive multimedia tools, many universities now offer credit bearing, fully accredited undergraduate and graduate degree programs via distance learning.

For more than a decade, Rochester Institute of Technology (RIT) in upstate New York has provided both liberal arts and technical distance-delivered education. As one of the early pioneers in distance learning, RIT distinguished itself by providing distance-delivered education to students located around the world. As an academic environment, RIT's student population is made up of the traditional undergraduate student 17 to 21 years in age, a large population of deaf students, graduate students, and adult learners who seek to enhance their skills or change careers.

In the beginning, distance learning was provided to the aforementioned population using videotaped lectures, textbooks, and other printed materials. Students registering for distance learning classes at RIT could do so by telephone or by visiting the campus. Once enrolled, course videotapes and other supporting materials (typically printed) were mailed to the student. In areas where there were a large concentration of distance learners, arrangements were made with local cable and TV stations to broadcast videotaped lectures at predetermined times.

While this initial model was an amazing success, many issues and challenges were identified. Videotaped and printed materials aged quickly. The cost associated with retaping videos and revising paper documents was escalating. Additionally, the cost of creating and distributing paper documents to students was high.

In addition to cost management, one of the chief concerns of those who participate in and design distance-delivered education is interactivity. Both students and teachers expressed concerns that the experience provided in the distance learning environment mirror that of the same courses taught on campus. By extending support facilities to include audio conferencing, Internet access, and tools such as e-mail, group conferencing, electronic reserves (where documents are stored as PDF files), and chat sessions, a new generation of distance learners have completed both graduate and undergraduate degree programs at RIT.

There is generalized agreement in the academic community that videoconferencing provides the most interactivity in distance-delivered education. However, the trade-offs between bandwidth, speed, and cost, make it difficult to provide real-time video solutions. Desktop videoconferencing is being explored as a possible candidate for increasing interactivity in distance-delivered content. However, limitations in resolution, the number of simultaneously attached users, and other factors, suggest that desktop videoconferencing in its current iteration is not suited to the task.

The virtual classroom provided by distance-delivered education holds the promise of not only providing anytime, anywhere education. More importantly, interconnected universities and campuses around the world can benefit from the interaction between the key thinkers and specialized resources located on these campuses. Once the bandwidth problem—the ability to provide high-resolution real-time videoconferencing in an affordable fashion—has been resolved, the use of other IT applications such as expert systems, virtual reality, and simulation, will become more commonplace in the virtual classroom of the future.

Chapter 9

Managing Network Resources

STRUCTURED APPROACH TO NETWORK MANAGEMENT

Regardless of the technology deployed or product or services produced, the success of an organization is based on the extent to which the company understands and marshals its policies, information resources, and individuals strategically. These factors are so fundamental, they are often taken for granted. Perhaps the most important, and oftentimes overlooked of these factors is the importance of human resources in information network management.

As information systems (and their underlying networks) have grown in complexity and importance to organizations, the corresponding expertise or knowledge necessary to interact successfully with these information resources has changed radically. The divestiture of AT&T placed many corporations in the role of managing their information destinies. During the 1980s to mid-1990, most organizations managed voice, data, imaging, and other data as functionally separate networks, each with associated administrative components.

Traditional network management strategies such as data processing and management information systems (MIS) focused on the internal functional and operational aspects of network management. Both approaches were characterized as monolithic entities that have existed behind a cloud of mysticism. During this phase network management activities, though highly organized, were at best reactive to changes required in technical areas such as capacity planning, performance management, and problem management.

The proliferation of computer technology, developments in Very Large Scale Integration (VLSI), declining costs in analog and digital transmission facilities, and a staggering array of Customer Premises Equipment (CPE) offerings have had the cumulative impact of decentralizing and demystifying the network management process. In order to move companies to a more proactive posture in managing these resources, a break from traditional Data Processing (DP) and MIS was indicated.

Developing and implementing business plans, which include an analysis of information technology and applications, is a relatively new trend. By matching information technology to the organization rather than the organization to the technology, modern enterprise has embarked on a path that emphasizes the strategic integration of information technology.

While this looks good on paper, many companies are faced with managing information resources that often rival their primary line of business. A complex hybridization of analog and digital, wired and wireless facilities, multiple network computing environments (where peer-to-peer, client/server, and other information systems coexist), and the ever-looming threat of technological obsolescence have caused many companies to shudder at the very thought of developing a structured approach to network management.

While the task of managing network resources in the Information Age seems daunting, a number of approaches have emerged to provide a template for developing a winning network management strategy. Before these approaches are explored, let us look at some of the basic functions and issues that must be addressed when managing network resources.

Network evolution is an ongoing process facilitated by careful network management. Network management as a process is comprised of the following interrelated processes:

- Integrated systems planning and control – an integrated approach to network management where information technology and its associated applications are seen as an integral part of the enterprise strategic plan.

- Performance management and capacity planning – ensures that all aspects of the network are optimized to meet existing and future needs.

- Configuration management – keeps track of all elements of the network such as transmission infrastructure, hardware, software, and who has functional responsibility for various segments of the network.
- Problem management – is concerned with the detection, prevention, and prediction of network outages, faults, or other disruptive events.
- Change management – focuses on the integration of information technology into existing policies, business practices, and job functions.
- Security management – protection of data and the network from unauthorized access, manipulation and destruction, and other potentially harmful and often illegal activities.

Each of the elements in this structured approach to managing network resources are highly interdependent. Performance management and capacity planning are dependent on configuration management so the status and location of network components are known. Historical data from problem management guides network designers in selecting the appropriate mix of resources for network upgrades.

Change management ensures that proposed network changes are integrated smoothly into the organizational fabric. By adopting a user-centered approach as part of the overall migration strategy, users are able to adapt quickly to newly available resources, resulting potentially in an increase in overall productivity. Finally, a clear understanding of the scope and power of enterprise-wide networking capabilities enables a company to strategically deploy these resources as part of an integrated business strategy.

Network management is made less stressful by the availability of GUI-based network management tools that collect, summarize, and model data for high-speed interactive multimedia networks. In the past, these packages and tools were specialized or specific to a particular type of data traffic such as voice, data, or video. Using broadband digital network as the backbone, a diverse range of data may be carried over shared facilities creating a focal point for developing an integrated network management strategy.

MANAGING NETWORK GROWTH

Most network managers will liken their job to trying to maintain their balance on top of a large rolling ball headed toward the edge of a precipice. This is not an exaggeration considering the need to maintain and maximize current investment in network resources in the face of obsolescence, global competition, and a staggering number of network offerings and solutions.

Managing network growth is just another part of overall network management. The current and estimated needs of an organization must be weighed against cost, compatibility, and interoperability. The task is not easy because the organization and its associated processes become more sophisticated as a result of integrating these network tools. This in turn drives demand for more information resources making network evolution a perpetual process.

Few organizations have developed or followed (if available) any structured approach to network planning and design. Seldom has this process lent itself to easy analysis. Gaining an understanding of the complex interaction of international standards, domestic regulatory and trade issues, long-term planning and network economics are necessary. The primary factors that must be considered in developing a cohesive network migration strategy include, but are not limited to, the following:

1. Modeling of the current network and traffic patterns
2. Modeling of the proposed network
3. Development of a cyclic model of traffic growth—capacity planning
4. Technology assessment
5. Identification of the appropriate migration strategy
6. Identification of a range of network alternatives
7. The selection of the most appropriate network evolutionary path

The strategic benefits of developing a cohesive network migration strategy include:

- Matching information technology with stated business goals and objectives as part of the planning process
- Identification of those critical elements of corporate information resources that will aid an organization in meeting its strategic objectives
- A better understanding of the interaction of corporate processes, human factors, and technical considerations
- Evolving the company to a state of competitive maturity
- Escalating the company's status from a reactive mode to a proactive state of preparedness
- The development of a flexible, long-term network strategy

In order to ensure that the proposed network changes are integrated seamlessly into an organization, a careful analysis of the operational considerations must be undertaken. The identification of operational factors must look not only at existing technology, applications, and infrastructure, but include a clear picture of workflows, the flow of information (formal and informal), and other organizational interdependencies.

Unfortunately, the reality of running a modern enterprise often makes this seemingly time-consuming, structured approach more of an impediment than an enabling process. Adopting an interactive, user-centered approach (fancy way of saying get the users on board from the start) to network design is well worth the time and effort to conduct. Once completed, the database and documentation will provide a ready base for future network migration. Table 9–1 summarizes some of the activities in the design process.

Table 9-1 The Systems Development Life Cycle	
Phase	**Focus**
Investigation	True nature of the problem or requirement Scope of the requirement or problem Objectives
Systems Analysis	Data gathering: Written documentation Interviews Questionnaires Observation Sampling Data analysis: Charts Tables System requirements
Systems Design	Alternatives Output
	Input Files Processing Controls Backup
Systems Development	Programming Testing
Implementation	Training Equipment conversion File conversion Systems conversion Auditing Evaluation Maintenance

NETWORK SECURITY

Network computing has become so ubiquitous that we often take for granted the risks associated with having 24-hour-a-day, 7-day-a-week access to applications online banking, Internet shopping, and the global availability of automatic teller machines (ATMs). We stare with amazement as we withdraw, deposit, or manipulate funds from our accounts located in our country of origin. Standing in an airport in France we are able to gain access to our accounts due to technological innovation and a complex series of agreements that foster this degree of interoperability and access. Each ATM placed in villages, retail outlets, and other public access points represents a potential point of unauthorized or unwanted entry to the financial community's information resources.

Perhaps the biggest culprit in the network breaches is the very feature that makes network computing possible—standardization. Standardization ensures that data and transactions can move seamlessly across a vast variety of disparate network resources. However, standardization also limits the number of choices and attempts that a hacker or other unwanted individual must make before access is gained to network resources.

For example, many telephone systems or PBXs support network management capabilities to allow dial interaction with a system. This enables network problems to be isolated, diagnosed, and repaired regardless of its location. Access to PBX supervisory functions is typically based on a four-digit personal identification number (PIN). Using a random number generator, a hacker can break into any voice or data network using a standard touch tone telephone.

Volumes have been written on network security often confusing rather than helping those corporations interested in implementing security measures. The following section summarizes some of the considerations, definitions, and breaches that are associated with modern network computing.

Network vulnerability can occur at the following points:

- terminals, workstations, and other access devices
- automated log on procedures
- remote or dial capabilities to system resources
- advanced inward dialing features such as 800 numbers, WATS, and other special access numbers

Network breaches can take the following forms:

- unauthorized access to network resources
- unauthorized access to data resources

- destruction of network resources
- destruction of data resources
- unauthorized use of network services
- unauthorized use of computing resources
- unauthorized use of transmission facilities
- unauthorized copying of programs and data files

It seems that hacking has become a competitive sport. Each day we read headlines detailing how some major corporation's network was entered and abused. Unauthorized use of and access to network computing resources is illegal. However, the rapid diffusion of information technology into almost every area of business and our daily lives make detection and enforcement of penalties problematic. While the AT&T case did much for educating the judicial system about the intricacies of information technology, many courts do not fully comprehend the magnitude of the damage caused by hackers. Hacking and other related activities are seen by their perpetrators and unfortunately the judicial system as victimless crimes.

Hacking into corporate and other information networks is far from a harmless, haphazard endeavor. Hackers have radio and TV broadcasts, underground networks, often piggybacking on the same networks that were hacked. Magazines and other popular media have escalated hacking to the status of super stardom complete with a cult following.

Hacking, viruses, and other illegal activities fall under the category of computer crime. Advances in secondary storage have facilitated the unbridled duplication and distribution of computer software. Telecommunication facilities used to support the virtual workplace often act as a welcome sign to unauthorized users. The size and complexity of many corporate networks make the detection of these harmful events difficult, if not impossible, and equally difficult to prevent.

Hacker Vocabulary and Techniques

Hacker – is a term generically associated with people who break into network computing resources or alter computer program source codes. While some hacker activities may be harmless, any unauthorized entry into a network and its associated resources is illegal. Hacker hotlines, magazines, dictionaries, and even Internet resources make it nearly impossible to stem the tide of this behavior.

Phreak – this is a term used to describe someone whose specialty is breaking into corporate telephone systems. Unauthorized entry into telephone systems is perhaps the most vulnerable link in network information resources. Phreakers typically perpetuate toll fraud by hacking into PBX systems as well as common carrier facilities to make illegal long distance telephone calls.

Computer Viruses

It is estimated that between five and seven new virus definitions are created each day. A computer virus is a program or code used to infect operating systems and software applications. Viruses typically cause destructive or other unwanted events on the system where they reside. The impact of viruses can range from little more than a simple prank to bringing a computer system to a grinding halt.

Viruses spread from one machine to another typically through the illicit copying of application programs. In networked environments, the impact of viruses have been devastating. The increased use of network computing resources not only provides a ready vehicle for infestation but is often the source of the virus. When we surf the Internet we pass through countless numbers of servers and other communications equipment in order to retrieve information or Web pages. Viruses may infect our systems through the programs we download, e-mail, and other Web-based resources.

Computer viruses can be grouped into three categories:

1. Macro viruses – infect application programs such as Microsoft Word and are generally little more than a nuisance.

2. Boot sector viruses – infect the executable files of an application program or operating system. They attach themselves to the boot sector of diskettes or the master boot record located on hard drives.

3. File infectors – are a type of virus that attach themselves to program files that have .exe or .com extensions. Other program extensions are also vulnerable to this type of virus.

Bomb – is the computer program that causes a network operating system or program operating system or network to crash.

Worm – is a type of computer virus or program that infects network resources by replicating itself. Worms typically propagate on hard drives, secondary storage, and memory having the cumulative impact of eventually crashing the system.

Trojan horse – is a computer virus or routine that associates itself with a valid application program or software. The Trojan horse is far from harmless. Oftentimes, computer passwords and other protected information are located and stolen. The Trojan horse is also used to alter application programs, making them easier to copy or access.

Data diddling – refers to the unauthorized and illegal alteration of data that resides on the computer network or as it is transmitted through the network or as it enters the network.

Piggybacking – can assume a number of forms in modern computer networks. In the computer environment, piggybacking allows unauthorized users to gain access to computing time and applications by using or by mimicking the identity of an authorized user. Voice mail systems are another popular venue for piggybacking. In this instance, unused voice mail boxes are secured by hackers and used to distribute authorization codes, credit card numbers, and other illegally obtained data.

Antivirus applications – provide a means of "inoculating" software resources infected by viruses. Antivirus software can detect a virus before it is loaded onto a system and subsequently a network. Antivirus programs not only detect viruses that might reside on a computer system, they also allow them to be deleted or inoculated.

Data Encryption

By translating all forms of data (voice, data, video, audio, and other interactive multimedia) into binary form, we are able (using digital transmission facilities) to encrypt these precious resources. Data encryption involves rendering data inaccessible or unusable to unauthorized users. This is achieved through the use of a cipher. The transmitting device sends secure e-mail or other forms of data that is encrypted using a given cipher. The receiving device decrypts or converts the encrypted data back into its original form. This is possible through the use of an algorithm or key used to unlock the cipher.

The proliferation of digital transmission has created a number of problems for law enforcement officials. By transmitting digital data over both wired and wireless media, illegal traffic and transactions go relatively unscathed from the watchful eye of the law. Additionally, the use of encryption products such as pretty good privacy (PGP) and more recently strong encryption, has stimulated a great deal of debate. Using these encryption technologies makes it virtually impossible to break or decode transmitted messages. The federal government has proposed the use of both a private and public key. The private key essentially unlocks the algorithm used to encrypt the message or data, allowing intervention by law enforcement officials.

While there are many techniques for encrypting data, perhaps one of the oldest and most widely used was developed by IBM Corporation and adopted by the Department of Defense (DOD). This standard is called data encryption standard (DES). DES works by using a private key. This key applies a 56-bit algorithm to each block of data that is 64 bits in length. It is possible to generate 72 quadrillion distinct randomly generated keys when using DES.

Biometrics

Biometrics is a form of system and data protection that authenticates the identity of authorized users on the basis of a unique physical characteristic. Voice recognition, retinal scanning, fingerprinting, or the analysis of some other unique anatomical feature is used to prevent unauthorized access to system resources.

Firewalls

The popularity of the Internet has created innumerable security problems for attached networks and provides a nearly untraceable conduit for computer crime. A firewall is a type of gateway server software used by companies to restrict access to, or the transfer of, certain types of information to unrecognized Internet domains or addresses.

Some security tips follow:

- Do not use names, nicknames, or other diminutives as passwords.
- Avoid using the same password for all of the resources you have access to.
- Change passwords frequently.
- Do not allow others to use your password or PIN.
- Turn off or deactivate unused resources or devices attached to the network.
- Conduct a periodic audit for unusual activity or patterns.
- Educate users to potential hazards.
- Develop and enforce acceptable use guidelines.
- Break into your own system to identify “open doors.”
- Install and use antivirus software.
- Secure wastes by shredding, etc.

Chapter 10

What Every User Should Know

TELECOMMUNICATIONS ACT OF 1996

As we approach the turn of the century, there are a number of challenges and opportunities facing the telecommunications, electronic communications, and information technology markets. To explore some of these issues, let us begin with an overview of telecommunications regulation in the United States. Given the complex interaction of technologies and applications over the years, telecommunications policy in the United States has been characterized by a number of futile attempts to delineate the infrastructure from the services it transports.

Understanding the strategic role that telecommunications played in trade, national security, and the economy, the United States government took steps in 1934 to protect its telecommunications resources. The first comprehensive regulation of the telecommunications industry was in the form of the Communications Act of 1934. The Communications Act of 1934 empowered Congress to oversee all activities pertaining to the regulation of telecommunications in the United States, and established the Federal Communications Commission (FCC) to regulate both federal and state telecommunications and the PUC/PSCs to provide state level oversight.

The Federal Communications Commission exercised oversight over interstate as well as local telecommunication traffic. Understand that in 1934, the public switched telephone network consisted primarily of wired and wireless facilities carrying voice traffic. As new products and services were added to the PSTN, the FCC found itself with a very difficult task. The provisions of the Communication Act of 1934 allowed for the creation of what was called "natural monopolies" within the local portion of the telecommunications industry.

Congress and the FCC thought that the needs of providing high-quality, ubiquitous telecommunications service were best served by having a single common carrier provide services within a geographic territory known as a local access and transport area (LATA). Those common carriers responsible for providing service within these LATAs were called *intra-exchange carriers* or *telcos*.

By connecting all of the LATAs throughout the United States using long distance links, inter-exchange carriers or long distance companies tied the PSTN together. Therefore, the role of regulating telecommunications in the United States was shared at the state level by entities called *public utility commissions* (PUCs) or *public service commissions* (PSCs) and the FCC. The tariffs or rates charged for local service were set at the state level through a complicated process, which required telcos to apply for rate changes through written and public hearings. By dividing responsibilities for managing the telecommunication industry between the federal and state governments, jurisdictional conflicts were not uncommon. Table 10-1 highlights some of the key issues and activities that helped to shape our present information environment.

From 1934 to 1981, AT&T dominated the telecommunication landscape in both the local and long distance markets. Consequently, there was little in the way of flexibility, variety, or competition available for consumers or businesses in terms of telecommunications equipment and services. The 1981 divestiture of AT&T opened the telecommunications market to fierce competition from both domestic and local telecommunications service providers and equipment vendors.

From 1984 until 1995, chaos characterized the telecommunication landscape. Many businesses found themselves for the first time confronted with the need to integrate services and technology from various vendors and of different architectures. In the meantime, policymakers struggled to protect and maintain order in the burgeoning telecommunications industry.

Many lessons were learned from the period between 1934 and 1984. Perhaps the most important issues centered around developing a national strategy for encouraging innovation while at the same time ensuring fairness and competition. Hence, what has become known as the *Telecommunications Act of 1996* was born. It is not very hard to understand the importance and magnitude of this Act. The following section summarizes some of the key issues that were addressed as part of this progressive legislation.

Table 10–1 Chronology of Telecommunications Milestones and Regulation

Date	Event
1835	Invention of the telegraph by Samuel Morse.
1844	Morse and Alfred Vail exchange first long distance message between Washington and Baltimore.
1850	Associated Press utilizes telegraphy for articles that appeared in nearly every daily American newspaper.
1866	Post Roads Act passed (to regulate telegraphy).
1874	Harmonic telegraph invented by Bell – allowed the same wire to transmit 30 – 40 messages simultaneously.
1876	Alexander Graham Bell received the patent for the analog voice network. Thomas Edison invented carbon transmitter – transmits speech over long distances.
1877	Bell Telephone founded.
1882	First manual switchboard was developed so that a centralized location could be used for connecting people by telephone.
1888	Interstate Commerce Commission (ICC) formed to regulate telephone service.
1889	Almon Strowger invented the step-by-step switching (Space Division) – lessened for direct manual connections between telephones.
1890	Electronic tabulating machine invented for census tabulation and calculation (computer precursor).
1892	New York to Chicago telephone line installed.
1895	Marconi perfected wireless telegraphy.
1896	Strowger invented telephone dial.
1900	American Telephone and Telegraph founded – created a monopoly by merging Bell Labs, Western Electric, Long Lines and operating companies.
1901	Trans-Atlantic radio communications service available.
1904	Korn developed telephotography – send photos via telephone or radio circuits.
1915	First transcontinental U.S. telephone line in use.
1918	Multiplexing was developed as a method of combining two or more voice channels on a single wire.

(Continued on next page)

Table 10–1 Chronology of Telecommunications Milestones and Regulation (continued)

Date	Event
1934	Communications Act of 1934 passed Legislated formation of the Federal Communications Commission (FCC). Public Utility Commission/Public Service Commission (PUC/PSC) chartered with interstate regulation of telephone service.
1935	Crossbar switch speeds up switching of voice between telephone companies.
1948	Concept of information theory developed by Claude Shannon in Mathematical Theory of Communications; incorporated a measure of limitations and possibilities of a communications channel.
Early 1950s	Microwave communications links established – provided high-volume communication over fairly long distances as long as the "line-of-sight" path for each transmitter was unobstructed; transistors developed during this decade.
1958	Data communications developed; front-end processors began to use voice lines to transmit data.
Early 1960s	"Picturephone" available from AT&T to transmit video pictures, but financial and bandwidth costs impeded feasibility of this option.
Mid-1960s	First tone-generating telephones developed, first electronic switching systems.
Late 1960s	ARPANET – first large-scale computer network developed for the Advanced Research Projects Agency (ARPA) part of the U.S. Department of Defense, was intended to link computer scientists at universities and other research institutions was introduced as a minor feature of the network – but quickly became popular.
1968	Caterfone connects mobile radio systems with telephone network – allowed companies other than AT&T to use AT&T lines.
1972	Microprocessor switch – provide call waiting, call forwarding, etc.
Early 1980s	Videoconferencing available (costs reduced and advances increased since 1960s), although costs were still high and image quality was poor.
1996	The Telecommunications Act of 1996 was enacted as the first comprehensive legislation to address competition, oversight, and other issues related to the telecommunications industry since the Communication Act of 1934.

- Defining the role of telecommunications and information technology in the medical and allied health fields
- Strategic role of information technology in national and international trade
- The impact of information technology on education in America
- Delineate the roles of private and public entities in shaping the telecommunications environment in the United States
- Stimulation of competition and fair business practices within the information technology industry and its associated markets
- The enhancement and protection of individual rights in the Information Age (quality of life and increased participation of the citizenry due to the use of IT in the political process)
- The redefinition of universal service in the United States

The Telecommunications Act of 1996 provided policy in the areas of information services, television and radio broadcasting, cable television, telecommunications services, and telecommunications equipment and products. In terms of information services, content became the primary focus of this bill. It is perhaps this portion of the bill that actually made the average citizen aware of its existence.

The Federal Communications Commission was given interim oversight over the Internet. In short, a moratorium was placed on all explicit, pornographic, controversial, and other objectionable activities or discussion. These activities conducted over the Internet or other information services were deemed illegal. This provision alone sparked unprecedented public protest and enumerable constitutional challenges. This aspect was later overturned.

In addition to issues related to profanity and the Internet, the Telecommunications Act also sought to address and delineate the rights and responsibilities of information service providers. The debate in this area focused on who would be held responsible for the commission or proliferation of unethical, immoral, or illegal activities using the Internet or other information services. Up until the Act, the courts had held the actual perpetrator, service provider, or both the perpetrator and service provider, liable for offensive or illegal activities. In short, there was no clear-cut precedence.

Traditional radio and television broadcasting had long been the favorite target of regulation in the United States. However, to protect these national resources, in the face of competition from other information technologies such as direct broadcast satellite, cable TV, and the Internet, these industries had been relatively unregulated. However, since a few cross-media ownership restrictions remained, the Federal Communications Commission found it necessary to revisit this area of regulation.

The FCC determined that cross-media ownership was, in fact, a positive phenomenon within the TV and radio industries. The Act also sought to address the need for more spectrum allocations for providing new wireless services. While the television and radio industries benefitted from the provisions of the Telecommunications Act of 1996 an old nemesis came back to plague them. Over the years, much attention had been focused on the nature and content of TV and radio programming.

Violence, nudity, obscenity, and other related issues have always plagued the TV and radio environments. While many agreed that something needed to be done, they also acknowledged the constitutional protections relevant to written and spoken expression. The Telecommunications Act proposed the inclusion of a V or violence chip in all televisions manufactured after the enactment of the bill.

The cable TV environment had strenuously sought regulatory relief in the face of stiff competition from broadcast satellite and other computing services. At the heart of this debate was the potential entry of telecommunications providers into the cable TV industry and conversely, the entry of cable TV providers into the world of telephony.

Both the cable and telephony providers had vested interests in wanting to be on the other end of the single cable that would bring information and other content into American homes in the future. The Act provided the regulatory relief necessary to allow these entities to explore new markets. The cable TV environment did not go unscathed in terms of the regulation of content. The bill required the cable industry to scramble or encrypt all programming containing objectionable, explicit, or adult themes.

Telecommunications services and products are so closely intertwined that it is often difficult to mark the line of demarcation between them. The Telecommunications Act of 1996 provided much regulatory relief for the Bell operating companies (BOC). Based on Justice Green's Modification of Final Judgment (MFJ) as part of the divestiture of AT&T, the Bell operating companies were forbidden from manufacturing telephones and other related devices. The Telecommunications Act lifted this restriction.

Some of the cross-jurisdictional issues that prevailed in the area of long distance and local telephone services were ironed out as part of the bill. The bill also opened the door for increased competition within the local telecommunications services environment. Finally in the area of telecommunications services the bill provided for the entry of the BOCs into the long distance arena.

OWNERSHIP OF INFORMATION

Remember the story your parents used to tell you at bedtime? You know, the one about the princess who saved her kingdom from certain doom. When you grew up you remembered this happy tale and decided to write a short story for children. The book is a commercial hit and you are approached by a Broadway producer who would like to adapt the story for stage. Again another commercial success. The play is then made into a feature film complete with a princess action figure, cereal, toothpaste, and underwear contracts. Just when you are about to bask in the glow of commercial success, your attorney advises you that you have been named in a lawsuit for stealing the creative ideas of another.

While this scenario sounds a bit outrageous, each day American courts convene to adjudicate such cases. Add to this scenario the fact that online services and the Internet provide new venues for the distribution and expression of intellectual property and other creative works. The issue that is at the heart of this example focuses around ownership of information. Prior to popular interest in the Internet, other forms of electronic mass media raised the specter of ownership of information. In the guise of the public's right-to-know, information about the most private aspects of one's life might be exposed to the world.

Networking has only contributed to the difficulty in regulating information. The connection of computers around the globe fostered the transfer of information about individuals as well. Enhancements in database management and distributed data processing make it possible to gather the most intimate details of an individual's life with a single keystroke. What we have begun to understand is that we as individuals have scant little in the way of control over the accuracy, creation, and distribution of information about ourselves. Again, the presence of online services and the Internet increase the magnitude of this issue.

Another issue related to the ownership of information or what is legally referred to as *intellectual property* is the determination of when a creative work is no longer the property of its creator. Suppose you take an award-winning photograph. Of course you copyright the photograph, which entitles you to royalties for its use. You create a home page on the WWW as a means of extending your presence to a new generation of viewers. The photograph is downloaded and massaged using a popular photo finishing computer application. The colors were changed as well as many other elements of the original photograph. One night you are surfing the endless maze of TV channels and you come across a commercial, which has an image that reminds you of your photograph. Upon investigation you learn that your photograph was indeed used as the basis of the image contained in the TV commercial. In court the plaintiff argues that, yes, your photograph was the basis of the image. However, the photograph no longer represents or bears any resemblance to the original image.

Again, this scenario highlights issues related to the ownership of information. In this instance the information is a photograph. But what if this intellectual property were a song, video clips, computer programs, and other interactive multimedia content? Does the issue of ownership become any less problematic in the Information Age? To summarize some key issues or questions related to the ownership of information or intellectual property, the following list has been provided:

- Who is responsible for the accuracy of information?
- How can the individual determine the extent to which information about himself or herself has been created, distributed, and used?
- When we detect a breach or problem with intellectual property, how can we determine the breadth of its distribution or assess the damages caused?
- What remedies exist for the individual when intellectual property has been wrongfully or illegally used?

- If the Internet or an online service provider is used for the venue where some breach of intellectual property has occurred, should the ISP or the service provider be liable in any proceedings?

These are just a few of the issues that we must address as participants in the Information Age. Many suggest that the magnitude of this problem renders the issue academic. Others suggest that magnitude should not be used as an excuse for failing to protect the intellectual and creative property of individuals or organizations.

PRIVACY

Privacy is defined as the absence of publicity, being away from others, solitude, or seclusion. Perhaps no other issue has been the focus of such intense public scrutiny as privacy. Many would argue that privacy is a vanishing commodity in the Information Age. The widespread deployment of high-speed infrastructure, powerful computing resources, and a staggering array of databases makes privacy a difficult right to protect. This section addresses some of the salient issues related to the topic of privacy in the Information Age. While volumes have been written on this topic, just a few of the key issues are summarized in the following list:

- Data collection, storage, and access
- Privacy in workplace

    E-mail

    Phone calls and voice mail

- Protecting privacy – regulation and remedy

Technology has become a double-edged sword. Innovations in mass storage technology coupled with advanced database management tools have made data collection and integration a major element of modern enterprise. Powerful spreadsheets and database application programs allow an individual with a modest computer configuration to access, store, and manipulate vast amounts of information. For example, many records about us exist publicly. The white pages, property and other tax data, census data, the city directory, and other seemingly harmless files can become powerful deterrents to privacy. By storing the data from the aforementioned sources then submitting them to a cross-attribute search using any database program, it is possible to pinpoint key information about an individual that was obscured in the sheer volume of these public records.

Each day thousands of pieces of information are collected by commercial, government, and other organizations about individuals. What is ironic or even perhaps frightening is that we seldom know that this information is being gathered and stored. We have no idea who has access to this information, to what uses this information is being put, or the accuracy of that data. Credit reports have long been a source of controversy. Used as the basis for transactions such as purchasing a home, determination of creditworthiness, or in your application for a new apartment, few of us actually know what is contained in our individual credit reports. In the instance we are denied credit, the law provides for us to have access to our credit report free of charge. Under any other circumstances, credit bureaus charge a healthy fee to obtain a copy of our own credit history.

When we complete an application, obtain goods or services, or buy our food at our local supermarket, information about these transactions are automatically gathered and stored in some central database. Oftentimes, this information is re-packaged and sold without our knowledge. Add the Internet or other online services to this scenario and we begin to get an idea of the magnitude of this problem. One of the things that makes the Internet an attractive venue for communication is that it allows each of us to become publishers and share any information we choose with the rest of the world. In most cases, we have no idea who has visited our Web site, what information was taken from the Web site, and how that information is subsequently used.

While most attention regarding privacy has focused on the Internet, online services, and the workplace, perhaps the most overlooked aspect of privacy has to do with system integrity. Each day we use our telephones, cellular telephones, cordless telephones, and other technology to communicate the most intimate details of our lives. Sophisticated technology exists and is sold legally, which allows for the undetected invasion of our privacy. Information systems are another potential source for the invasion of individual and corporate privacy. Each minute millions of dollars in transactions are carried across networks that attach automatic teller machines to user accounts. Ironically, the high degree of standardization protocols, networking schemes, and authentication make it easy for privacy to be invaded.

Privacy in the workplace raises yet another set of issues to be addressed in the Information Age. E-mail has received a lot of attention in this regard. Unlike regular mail, e-mail does not fall into the same category as the letters we send through the U.S. postal system. While our letters are protected by law, e-mail, as a type of property, does not fall into the same category. It seems that the issue of ownership and the privacy of e-mail does not depend on who originates the message but rather who owns the resources used to create and transmit it. Other office technologies are at the center of similar debate. The use of the telephone or other voice technologies such as voice mail have also become part of the privacy debate. Does our employer have the right to monitor telephone calls, whether they are personal or business related? Does our employer have the right to listen to voice mail messages directed to us?

So what protection does an individual or a corporation have against loss of privacy? What remedies exist for us or corporate entities to redress privacy-related wrongs when we are able to detect them? A number of statutes and policies have emerged that attempt to address the issue of privacy. The Electronic Communications Privacy Act (ECPA) of 1986 was developed to protect private or confidential electronic communications. The ECPA evolved from a body of law related to electronic wire tapping. These earlier laws required law enforcement officials to obtain a judicial warrant before they could intercept any communication carried over a network or electronic transmission system.

In 1986 Congress soon recognized that these wire-tapping laws were not sufficient to the task of protecting us from the sophisticated range of electronic communications that has begun to emerge. The ECPA was extended to cover not only voice communications on the telephone, but any digital content such as images, sound, or any other multimedia data. Unauthorized eavesdropping is prohibited not only for law enforcement officials, but businesses and individuals are subject to prosecution for violation of this law. Given the complexity of the Information Age, it stands to reason that the ECPA must also evolve to cover all the potential abuses that could occur. Hence, the interception, unauthorized copying, or manipulation of any electronic message as it is transmitted or if it is stored is illegal under the ECPA.

Contrary to popular belief, neither the Constitution nor Bill of Rights explicitly guarantees or protects the right to individual privacy. However, the courts have upheld the right to privacy implied in the Bill of Rights and the Constitution. To add to the lack of jurisprudence in the area of privacy, many states now have some form of privacy law on the books; however, there is little in the way of consistency between these laws and their enforcement.

ETHICS IN THE INFORMATION AGE

Given the power of modern information technology, the issue of ethics comes to fore. Unlike other tangible resources, software, data, and the nontangible byproducts of the Information Age pose a number of new moral and ethical dilemmas. When polled the majority of people in a group indicated that they would return a wallet that they found. When asked if it was wrong to take paper or other tangible items on the office many answered yes. The same group was asked for a show of hands of those who used telephones at their place of business to make personal calls, used e-mail for personal messages, or voice mail for personal messages. Oddly enough almost everyone raised his or her hand.

What is it about information resources that make them a commodity or item that does not exist outside of cyberspace and therefore not real? What the previous scenario suggests is that while many individuals have strong taboos against stealing or lying, they seemingly have no reservations about using corporate resources for their own personal gain. In addition to ethics in the workplace, we must also consider how information technology is used as part of the decision-making or problem-solving process. Each day important decisions about creditworthiness, allocation of funds, admission to college, or other equally pivotal issues are made by computers. We must ask, "Are there decisions that computers should not make?" What amount of human accountability should coexist when computers are used in the problem-solving or decision-making process?

Consider the following scenarios. A patient is admitted to the hospital for a mild infection. His paper medical records indicate that he is allergic to penicillin. Recently, the hospital made a move to transfer all paper medical files to an electronic medical records system. Somehow during the transfer of this individual patient's records, his allergy to penicillin was not noted in the online medical database. During his stay in the hospital, the patient's temperature became elevated. Since the patient's vital signs were being monitored by sophisticated devices, the hospital staff administered a dose of penicillin. The patient dies as a result of the infusion of penicillin. Who is liable?

An applicant for a job is being interviewed by his prospective employer. He has completed a written application providing the standard information regarding employment history, academic experience, and related skills. Little does he know that the prospective employer has gathered other information such as a police record, credit history, medical records, and other personal information about the candidate. Based on this information, the candidate was deemed unsuitable and denied the position. Was the action of the employer wrong? Should the candidate have been made aware that this additional information would be considered as part of the screening process?

As a prank, a student takes a picture of his roommate and alters it so that the roommate appears to be dressed in a pink dress with heels. To add to the excitement, the student put a copy of the altered photo onto a home page on the Web. While wearing a pink dress is of course one's prerogative, what turned out to be an innocent prank backfired. Several years later, the roommate ran for public office. The picture resurfaced and was displayed on the front page of newspapers and in the advertising campaigns of his opponents. This is a clear case of defamation but who is to blame?

While these scenarios may seem exaggerated, they highlight some of the issues that are raised in our interaction with information technology. What remains to be explored are the acceptable and responsible uses of information technology. Social awareness must be raised regarding the potential hazards of technology and its byproducts.

Popular media is fond of pointing out the problems associated with computer crime or objectionable content. Perhaps more attention should be focused on the positive ways information technology keeps us connected to our global village and improves the quality of our lives.

Acronyms

ADSL	Asymmetric Digital Subscriber Line
AM	Amplitude Modulation
ANI	Automatic Number Identification
AOL	America Online
APD	Automatic Pharmaceutical Dispensing
ARPANET	Advanced Research Projects Agency NETwork
ASCII	American Standard Code for Information Interchange
ATM	Asynchronous Transfer Mode or Automatic Teller Machine
B-ISDN	Broadband ISDN
BRI	Basic Rate Interchange
CAS	Computer-Assisted Surgery
CDDI	Copper Distributed Data Interface
CDI	Compact Disk-Interactive
CDMA	Code Division Multiple Access
CD-R	Compact Disk-Recordable
CD-ROM	Compact Disk Read-Only Memory
CD-RW	Compact Disk-Rewriteable
CMOS	Complementary Metal Oxide Semiconductor
CPE	Customer Premises Equipment
CPU	Central Processing Unit
CSA	Cellular Service Area
CSCW	Computer Supported Collaborative/Cooperative Work
CSMA	Carrier Sense Multiple Access
CSMA/CA	Carrier Sense Multiple Access with Collision Avoidance
CSMA/CD	Carrier Sense Multiple Access with Collision Detection
DARPANET	Defense Advanced Research Projects Agency NETwork
DAT	Digital Audio Tape

DDP	Distributed Data Processing
DDS	Dataphone Digital Service or Digital Data Service or Digital Data Storage
DES	Data Encryption Standard
DNS	Domain Name Server
DOS	Disk Operating System
DP	Data Processing or Distributed Processing
DRAM	Dynamic Random-Access Memory
DSS	Decision Support System or Digital Signature Standard
DSU/CSU	Digital or Data Service Unit/Channel Service Unit
DVD	Digital Versatile (Video) Disk
EBCDIC	Extended Binary Coded Decimal Interchange Code
EFT	Electronic Funds Transfer
EMI	Electro-Magnetic Interference
E-1	European DS 1
EPROM	Erasable Programmable Read-Only Memory
FCC	Federal Communications Commission
FDDI	Fiber Distributed Data Interface
FDM	Frequency Division Multiplexing
FM	Frequency Modulation
FTP	File Transfer Program or File Transfer Protocol
GAN	Global Area Network
GEO	Geostationary
GUI	Graphical User Interface
HDSL	High Bit Rate Digital Subscriber Line
HTML	HyperText Markup Language
HTTP	HyperText Transport Protocol
IBM	International Business Machines
IDE	Integrated Drive Electronics
IDM	Integrated Data Management
IEEE	Institute of Electrical and Electronic Engineers

IMAP	Internet Message Access Protocol
I/O	Input/Output
ISDN	Integrated Services Digital Network
ISO	International Standards Organization
ISP	Internet Service Provider
JPEG	Joint Photographic Experts Group
LAN	Local Area Network
LED	Light-Emitting Diode
LEO	Low Earth Orbit
LOS	Line of Sight
MAC	Media Access Control
MACs	Moves, Adds, and Changes
MAN	Metropolitan Area Network
MPEG	Moving Pictures Experts Group
MRI	Magnetic Resonance Imaging
MTS	Mobile Telephone Service
NIC	Network Interface Card
NOS	Network Operating System
NSF	National Science Foundation
OSIRM	Open System Interconnection Reference Model
PBX	Private Branch Exchange
PCM	Pulse Code Modulation
PDF	Portable Data Format
PDN	Public Data Network
PGP	Pretty Good Privacy
POP3	Post Office Protocol 3
POTS	Plain Old Telephone Service
PPP	Point-to-Point Protocol
PRI	Primary Rate Service
PSTN	Public Switched Telephone Network
QIC	Quarter-Inch Cartridge

RAM	Random-Access Memory
RIP	Raster Image Processor
ROM	Read-Only Memory
RSA	Rivest-Shamir-Adleman
SCSI	Small Computer System Interface
SCSI-II	Small Computer System Interface Version – 2
SDH	Synchronous Digital Hierarchy
SET	Secured Electronic Transactions
SIMM	Single In-line Memory Module
SLIP	Serial Line Internet Protocol
SMDS	Switched Multimegabit Data Services
SMTP	Simple Mail Transfer Protocol
SONET	Synchronous Optical NETwork
SRAM	Static Random-Access Memory
SSL	Secure Sockets Layer
TCP/IP	Transmission Control Protocol/Internet Protocol
TDM	Time Division Multiplexing
TDMA	Time Division Multiple Access
T-1	T-Carrier Channel
URL	Uniform Resource Locator
VCR	Video Cassette Recorder
VDSL	Very High Bit Rate Digital Subscriber Line
VHS	Video Home System
VLSI	Very Large Scale Integration
WAN	Wide Area Network
WWW	Worldwide Web
WYSIWYG	What You See is What You Get

Glossary

100Base-FX – Provides 100 Mbps data transmission over two-strand fiber-optical cable.

100Base-T – By using more efficient network access protocols and faster transmission facilities, 100Base-T, also known as fast EtherNet, provides user access to network facilities of 100 Mbps over copper twisted pair.

100Base-T4 – Provides 100 Mbps to over four pairs of unshielded twisted pair (UTP)

100Base-TX – Provides 100 Mbps data transmission over data grade twisted pair.

10Base-5 – Specifies the use of a heavier gauge coaxial cable where each span supports 10 Mbps traffic up to 500 meters.

10Base-2 – Specifies the use of coaxial cable where each span supports 10 Mbps traffic up to 185 meters.

10Base-F – Specifies the use of fiber-optic cable to provide 10 Mbps data transmission up to 2,500 meters.

10Base-T – Perhaps the most widely implemented EtherNet standard. Using copper twisted pair, 10Base-T provides users with access to data rates up to 10 Mbps.

A

Absorption – Adverse weather conditions such as fog, rain, and snow can interfere with various forms of wireless transmission. The strength of the signal is distorted as it is absorbed by atmospheric conditions.

American Standard Code for Information Interchange (ASCII) – A 7- or 8-bit standard used to aggregate data bits fo transmission. ASCII is the most commonly used code set for microcomputer communications.

Amplitude – A characteristic of an analog signal that refers to its associated strength, intensity, or voltage. Amplitude is often measured in decibels (db).

Amplitude modulation (AM) – One of the earliest modulation techniques. This is the same modulation technique used to carry music over AM frequencies or on the radio. AM varies the height or intensity of the signal in order to represent transmitted user data.

Analog – A continuously varying or oscillating signal used to transmit data.

Anonymity – The need to maintain the confidentiality of all parties involved a given transaction. They include the seller, buyer, and any financial organization involved in the transfer of funds related to the transaction.

Antivirus applications – Provide a means of "inoculating" software resources infected by viruses. Antivirus software can detect a virus before it is loaded onto a system and subsequently a network. Antivirus programs not only detect viruses that might reside on a computer system, they also allow them to be deleted or inoculated.

Arithmetic logical unit (ALU) – Executes program instructions based on a number of logical operations such as and, or, not, less than, >, and equal to. Almost all program instructions can be executed as logical or arithmetic functions.

Asymmetric Digital Subscriber Line (ADSL) – Represents a new way of allocating available frequency using the subscriber loop. By dividing the bandwidth into three channels, one channel can be used to carry user voice signals. A second channel provides for transmission from the user (upstream) at the rate of 640 Kbps to a receiving entity. The third channel supports variable rates from 16 Kbps to 9 Mbps for downstream data transfer.

Asynchronous communications- Start and stop bits are added to the data stream to help distinguish the beginning and end of the transmission and the packets that are part of a given transmission and to maintain flow.

Asynchronous Transfer Mode (ATM) – A high-speed, digital network service defined as part of the emerging Broadband ISDN (BISDN) standard. As an emerging digital transmission technology, ATM uses fixed cells that are 53 bytes long. The first 5 octets or bytes are used for the header with the remaining 48 used to transport user data.

Attenuation – This is the tendency of signal strength to diminish as a function of distance. As signal travels across a link, its power or amplitude weakens. Wireless communications are especially susceptible to attenuation.

Augmented reality – The Internet or attached medical Intranets are being used extensively to provide medical education and training. Augmented reality allows training institutions to be connected to hospitals and other active medical environments. Essentially, it is virtual reality applications designed specifically for the medical environment.

Authentication – The ability to determine access and authority to conduct a particular transaction. The goal is to be sure that the parties involved are those who are authorized and legally recognized.

Automated pharmaceutical dispensing (APD) – A number of technologies expand medical intervention by dispensing medication based on predetermined criteria. APD assists medical practitioners in the calibration, preparation, measurement, and distribution of prescription medicine.

Automatic number identification (ANI) – Similar to caller ID. ANI is a service available to business clients, which provides comprehensive, customizable data about the calling party.

Availability and reliability – Availability refers to the period of time that the network is accessible whereas reliability is concerned with the integrity of transactions and applications available or travelling the network.

B

Basic rate interface (BRI) – The ISDN BRI supports data transport up to 192,000 bps by combining two B channels with one D channel.

Bearer channels (B channels) – Capable of transporting both circuit switched or packet switched traffic at 56,000/64,000 bps in the ISDN environment.

Binary – A two-state scheme where 0s and 1s are used to represent data and control characters.

Biometrics – A form of system and data protection that authenticates the identity of authorized users on the basis of unique physical characteristics. Voice recognition, retinal scanning, fingerprinting, or the analysis of some other unique anatomical feature is used to prevent unauthorized access to system resources.

Bit (binary digit) – The most fundamental unit for transmission or manipulation by a computer.

Bomb – The computer program that causes a network operating system or program operating system or network to crash.

Boot sector viruses – Infect the executable files of an application program or operating system. They attach themselves to the boot sector of diskettes or the master boot record located on hard drives.

Bus – In this topology, stations or nodes are attached to the network in a linear fashion.

Bus width – Refers to the size of the electrical pathways that connect the various subsystems of the computer to the CPU. Bus width is typically expressed 8-bit, 16-bit, 32-bit, and 64-bit. With a larger bus width, larger or more instructions and data can be sent to the CPU for processing.

Byte – A group of 8 bits for transmission.

C

Call conferencing – The ability to interconnect more than one party during an active telephone session. Usually requires more than one line and can be used in some areas in conjunction with call waiting.

Call forwarding – The ability to toggle with ease between wired, cordless, and cellular phones around the world. Callers can be switched to voice mail applications or an answering machine rather than encountering a busy signal, answering machine, or no answer.

Call return – Allows the subscriber to return the last call received on a given line. Oftentimes, callers will not leave messages in a voice mail system or on an answering machine.

Call trace – This feature allows law enforcement agencies to track and apprehend obnoxious or obscene callers.

Call waiting – Subscriber is notified of an inbound call attempt and has the option of placing the active session "on hold" while answering the new call. Recent enhancements allow caller ID to work in conjunction with this feature so the identity of the second caller is known.

Caller identification or automatic number identification – Caller ID is a service available to residential customers. It allows the called party to determine the identity (name and number) of the calling party. Special services or devices allow calls to be screened and forwarded, not accepted or disconnected based on subscriber requirements.

Carrier sense multiple access (CSMA) – A type of LAN access method where permission to transmit is granted to a station after the medium is checked to see if it is clear.

Carrier sense multiple access with collision avoidance (CSMA/CA) – In addition to incorporating the features found in CSMA/CD, this access method listens before, during, and after attempted transmission.

Carrier sense multiple access with collision detection (CSMA/CD) – As with CSMA the network is checked to see if it is clear. If two devices attempt to access the LAN at the same time, this access method will detect the collision and adjust subsequent attempts by these devices so as to minimize another potential collision.

Circuit switched network – Optimized for voice but provides connectivity and transport for other types of traffic modems and other devices.

Circuit switching – A switching system that completes a dedicated transmission path from sender to receiver at the time of transmission.

Client – In this network arrangement, the client is that network entity that requests computing services from other servers in the network. In short, the client requests computing services from other network-based software resources.

Clock – There are a number of internal or system clocks used in the computer environment. The CPU clock uses a quartz crystal to synchronize the processing of instructions per some given time interval.

Clock doubling – Technique used by experienced computer users where parameters changed in the configuration setup contained in the CMOS.

Clock pulse – Refers to the signal generated by the internal CPU clock. These pulses are continuous and correspond precisely to changes in voltage.

Clock speed – A metric of the number of pulses generated from a quartz crystal; used to provide timing for the CPU. CPU speed is typically expressed in MHz; i.e., 300 MHz Pentium II processor.

Code division multiple access (CDMA) – Also referred to as spread spectrum. Code sequences are used to provide multiple channels within the same broad of channel. Each channel is designated as PN1, PN2, and so on.

Code set – Standardized mechanism for grouping bits into a form suitable for transmission.

Collection of funds – Internet "funny money" or digital cash are among the many ways that customers and enterprises can pay for the goods and services exchanged. While credit card transactions make up the bulk of Internet commerce, new technologies and techniques are being explored to safeguard against theft and other forms of electronic fraud. Another issue that remains problematic are the remedies that are available to both consumers and enterprises in terms of refunds or collection of funds for items or services received.

Compact disk read-only memory (CD-ROM) – CD-ROM and its sister technologies offer the benefits of mass storage (600 MB+), portability, and competitive pricing. CD-ROMs are rapidly replacing floppy disks as the medium of choice for software distribution. 24X speed and higher CD-ROM drives provide high-speed access to photos, images, and software applications.

Compact disk-recordable (CD-R) – Drives that allow users to "burn" their own disks for in-house or external use. Adding record capability made CD-R an attractive optical storage alternative.

Computer-assisted surgery (CAS) – The widespread availability of high-speed, digital transmission facilities, advances in imaging techniques, and network computing have introduced a number of new tools into the surgical environment. Interactive real-time data in the form of x-rays, microscopy, and MRI scans can be transmitted in real-time to surgical experts and practitioners around the world.

Computer supported collaborative work (CSCW) – Has emerged as one of the new paradigms for the modern workplace. Technologies such as local area networks, software office suites, e-mail, and other forms of messaging allow coworkers to concurrently access and manipulate shared information resources. Groupware is a name that is given to the collaborative tools that foster CSCW.

Configuration management – Keeps track of all elements of the network such as transmission infrastructure, hardware, software, and who has functional responsibility for various segments of the network.

Control unit – The element of the CPU that regulates the flow data and instructions that will be processed by the CPU.

Cylinders – All of the tracks on the same disk surface of a hard drive.

D

Data diddling – Refers to the unauthorized and illegal alteration of data that resides on the computer network or as it is transmitted through the network or as it enters the network.

Data encryption- Involves rendering data inaccessible or unusable to unauthorized users. This is achieved through the use of a cipher. By translating all forms of data (voice, data, video, audio, and other interactive multimedia) into binary form, we are able (using digital transmission facilities) to encrypt these precious resources.

Data integrity – Concerned and closely related to issues of authentication and security. The goal is to ensure that all network-based transactions are conducted over secure facilities, that the data contained in the transactions is unaltered, and funds for such transactions are transferred in a timely and accurate manner.

Data transfer – Movement of data and instructions to and from the hard drive.

Delta channels (D channels) – Transport ISDN signalling or low-speed data at 16/64 Kbps.

Demodulation – The process of recovering modulated user data and control information from an analog line.

Digital – A discrete signal used as a means of transmitting data. All user and control data are translated into binary "0"s and "1"s.

Digital audio tape (DAT) – Represents an enhancement in sequential magnetic media. Used as the standard for professional recording, DAT never took off in the consumer market. By using a helical arrangement or digital data storage (DDS), the current generation of DAT technology can hold between 2 and 24 GB of computer data. Using DDS-3 DATs can be formatted to hold as much as 24 GB of data with transfer rates of up to 2 MB per second. DAT cartridges come in a number of formats such as 1/2 inch, QIC (quarter-inch cartridge), 8 mm and 4 mm.

Digital certificates – Another type of popular security method used to protect documents and other transactions on the WWW. Essentially, a certificate is a password-protected, encrypted file that contains information related to both public and private keys, identification of the user or valid users, the name of the certification authority, and the period for which the digital certificate is valid. A certification authority acts as a third party to ensure that the public key is authentic.

Digital signatures – In the real world authentication for retail, catalog, or other transactions is provided by an individual signature. This same concept has made its way into electronic commerce as a means of up authenticating, ensuring integrity and confidentiality of transactions. Digital signatures are unique random numbers that are created through a process called *hashing*. This uniquely created number is encrypted using a private key and is attached to the original transaction.

Digital videodisk (DVD) (more recently dubbed digital versatile disk) – A new optical disk technology that many predict will replace magnetic and optical alternatives such as VHS VCRs, CD-R, CD-ROM, and other forms of portable storage media. When compared to existing technologies such as CD-ROM, DVD not only offers increased storage capacity (17 MB) but it offers record capability on both sides of the medium. Current DVD standards support backward compatibility to CDI, photo CD, CD-R, CD-ROM, and other hybrid CD-based technologies.

Disk access time – A combination of seek time, rotational delay, head switching time, and data transfer. Data access time is measured in milliseconds.

E

Encryption – Involves rendering Internet transactions unreadable or otherwise inaccessible to unauthorized parties. In the area of Internet commerce, algorithms such as Rivest, Shamir, Adleman, encryption (RSA), pretty good privacy (PGP) used a combination of public and private keys, which are used to encrypt and decrypt Internet transactions. Remember, keys are specialized algorithms used to encrypt and decrypt data.

EtherNet – Jointly developed by DEC Corp., Intel, and Xerox Corp., EtherNet has emerged as the premier standard for local area networking. Based on the IEEE 802.3 standard for local networking, EtherNet is capable of providing data transmission rates of up to 10 Mbps to as many as 1,024 attached stations. EtherNet uses CSMA/CD as its access method. EtherNet's popularity is perhaps attributable to its flexibility across a variety of transmission media such as wireless, fiber-optic, copper tested pair, and coaxial cable.

E-time – The ALU accepts the decoded instructions from the control unit and performs the last two steps of the machine cycle. The steps are:

1. Executes the instruction (adds, subtracts, multiplies, divides, or compares)
2. Stores results of the process or execution main memory or registers

Extended Binary Coded Decimal Interchange Code (EBCDIC) – Developed by IBM for use in its mainframe computing environment.

F

Fiber distributed data interface (FDDI) – Pronounced FID-DEE. FDDI is based on the ANSI X3-T9 standard. FDDI is implemented In two versions – a fiber-optic version provides 100 Mbps transfer of data up to 124 miles. Using copper twisted pair 100 Mbps transmission is provided as CDDI. CDDI stands for copper distributed data interface; in this case, FDDI is implemented using UTP instead of fiber-optical cable.

File infectors – This type of virus attach themselves to program files that have .exe or .com extensions. Other program extensions are also vulnerable to this type of virus.

File server – A dedicated workstation whose responsibility is to manage shared operating functions and application software among connected entities. In addition to managing shared software resources, the file server also mediates access to physical transmission media and any shared peripheral devices attached to the network.

File transfer protocol (FTP) – Provides anonymous access to the resources on a specified FTP server.

Firewall – A type of gateway server software used by companies to restrict access to or the transfer of certain types of information to unrecognized Internet domains or addresses.

Frame relay – A more direct descendant of X.25 packet switching technology and provides for data-only support.

Frequency – A component of an analog signal that refers to the number of complete cycles or oscillations made by an analog signal over some time interval or complete signal changes an analog signal assumes in a given time element. Hence, the number of hertz (Hz) per second forms the basic unit of measurement.

Frequency division multiple access (FDMA) – Each user is assigned to a single frequency or signal among a band of available frequencies.

Frequency division multiple access/time division multiple access (FDMA/TDMA) – A hybrid combination of FDMA/TDMA is currently being used. The available bandwidth is divided into separate channels. Each channel is then divided into multiple time slots.

Frequency division multiplexing (FDM) – A line-sharing technique where the available bandwidth can be divided into separate frequencies separated by guardbands. In this instance, the data from various user devices is placed on a distinct range of frequencies.

Frequency modulation (FM) – Offers a better alternative as a modulation technique. By holding the amplitude and phase constant, user data is placed onto a line by varying the frequency in accordance to changes in the user data.

Full duplex transmission (FDX) – Provides for simultaneous, bidirectional data traffic. Four wires are used. Two can be used for transmitting and two to receive data simultaneously. As a result, more data can be supported at much higher rates.

G

Gigabit LAN – Transmission rates close to one billion bps over local area network technology.

Guided – Guided or wired facilities make up the majority of the infrastructure in the public switched telephone network and most private networks. Over the past 100 years, guided transmission facilities have evolved from copper to ultrapure glass or fiber optics. The three media that are commonly used to implement wired facilities are copper twisted pair, coaxial cable, and fiber-optic cable.

H

Hacker – A term generically associated with people who break into network computing resources or alter computer program source code. While some hacker activities may be harmless, any unauthorized entry into a network and its associated resources is illegal. Hacker hotlines, magazines, dictionaries, and even Internet resources make it nearly impossible to stem the tide of this behavior.

Half duplex (HDX) – In HDX transmission two wires are typically used: one to transmit and one to receive. As data is transmitted over one path, the other path remains idle waiting for the signal to change direction or perform what is called *line turnaround.*

H channels – Provide support for broadband data rates (384, 1536, or 1920 Kbps).

Head switching – The selection between the read head when transferring data from the hard drive or write head for transferring data to the hard drive.

H11 – In ISDN a single digital channel with a rated data capacity of 1536 Kbps.

Hertz (Hz) – A measurement of the number of signal changes or cycles completed as an analog signal is transmitted.

High Bit Rate Digital Subscriber Line (HDSL) – Provides for broadband digital transmission. By implementing HDSL, the local loop could provide data rates of 1.54 million bps or 2.048 Mbps. Since HDSL is full duplex, users have access to transfer rates between 1.54 or 2.048 million bps in either direction upstream and downstream.

H0 – In ISDN multiple H0 channels are used to provide data rates of 1.544 Mbps using three HO channels and one D channel or 2.068 Mbps connectivity using five HO and one D channels.

H12- In ISDN one 1920 Kbps channel is combined with one D channel.

Hybrid or mesh – The combination of bus, ring, star, and tree topologies within the same network.

I

Infrastructure – Refers to the complex mix of wired and wireless facilities used as the backbone or transport mechanism for a network.

Integrated Drive Electronics (IDE) – A type of hard drive that contains a built-in controller. The IDE drive is connected to the IDE host adapter with a flat ribbon cable. High-speed integrated host adapters can support multiple floppy drives, multiple hard drives, and other types of IDE input/output devices.

Integrated systems planning and control – An integrated approach to network management where information technology and its associated applications are seen as an integral part of the enterprise's strategic plan.

Internet Message Access Protocol (IMAP) – Provides the user with more flexible options for reading e-mail. Using IMAP, the user can download his or her e-mail, read it, and decide whether to delete it from the server.

Internet Service Providers (ISPs) – Provide users with either dial/dial-up or dedicated access to the Internet. Many commercial organizations use both dedicated and dial-up access to the Internet. Inexpensive Internet access is provided to noncommercial users through dial connections in the form of serial line Internet protocol (SLIP) or point-to-point (PPP).

Interoperability – Refers to the ability to connect to and the degree of compatibility necessary to support the exchange of information across networks of different architectures.

I-time – The control unit is that element of the central processing unit that initiates the first two steps of the machine cycle. The two steps that make up I-time are:

1. Fetch the next program instruction or data.
2. Decode the instruction or data.

M

Machine cycle -The CPU processes information in a series of operations known as the machine cycle. The machine cycle is composed of two constituent cycles: instruction time (I time) and execution time (E time).

Macro viruses – Infect application programs such as Microsoft Word and are generally little more than a nuisance.

Main memory – Also called primary storage or sometimes registers. This area of memory on the CPU is used to temporarily store the results of the various steps used to execute program instructions.

Master/Slave – In the master/slave network computing model, a mainframe or powerful minicomputer acts as the primary computing resource. Secondary computers and workstations attach to this central machine via dedicated or switched telecommunications facilities.

Megaflops – In minicomputers and mainframes megaflops refers to 1 million floating point operations or instructions per second.

Message switched network – Messages are sent (data or voice) as complete units through the network. Typically not real time. E-mail, voice mail, and other messages are stored and then forwarded once requested by the user.

Millions of instructions per second (MIPS) – Perhaps a more meaningful metric of system speed for personal computers.

Mobile Telephone Switching Office (MTSO) – The control center of the cellular network. Large, intelligent switches similar in function to those used in wired networks perform a complex series of processes that coordinate call setup, routing, authentication, maintenance, and termination.

Modulation – The process of placing user data and control information onto an analog line.

Modulator/demodulator (modem) – Used to superimpose user data and control information onto an analog facility and then recover both at the receiving destination. Modems are used to place video, text, and other multimedia data onto analog lines.

Multipath fading – As the signal leaves the microwave dish, it fans out causing a portion of the signal to be reflected back into the originating source. These wireless signals are reflected from the ionosphere or flat surfaces on the earth causing signal degradation and errors.

Multiplexing – A telecommunications transmission technique where the signals and data from many users are placed on the same transmission path.

N

Network – Refers to the integrated collection of hardware, software, and telecommunication facilities used to connect geographically dispersed resources.

Network interface card (NIC) – In order to attach the various nodes and stations to the LAN, a network interface card or NIC is required. The network interface card may attach a computer or printer to the LAN physically when guided facilities are used. In the instance where wireless facilities are used, network interface cards will have an antenna or equivalent sensor to receive the RF signals.

Network operating system (NOS) – A specialized operating system (OS) is required to implement the functions of a standard local area network. These functions include routing, addressing schemes, media access, and other specialized functions necessary to support a given environment.

Node – Refers to switching, routing, and other intelligent devices that make up a network. These devices may include computers, workstations, and specialized telecommunication devices such as front-end processors, modems, multiplexers, and other intelligent networking devices.

P

Packet switched network – An X.25 network that breaks data into information packets for transmission through the network; store and forward/real time.

Paging – A wireless component of POTS provides for digital and alphanumeric paging. Paging services may be implemented by the common carrier or the customer may manage his or her own paging network complete with base station.

Peer-to-peer – Peer-to-peer networking evolved solving many of the frailties associated with the master/slave model of computing. By distributing both computing and data resources among geographically dispersed nodes, a high degree of availability and recoverability characteristic of networks built on this computing model. In peer-to-peer computing, all attached resources are capable of performing the same functions.

Performance management and capacity planning – Ensures that all aspects of the network are optimized to meet existing and future needs.

Phase – Refers to the stage or point of a signal as part of some recurring sequence; in this case, the oscillation of an analog signal over time. Phase is typically measured in degrees (°).

Phreak – A term used to describe someone whose specialty is breaking into corporate telephone systems. Unauthorized entry into telephone systems is perhaps the most vulnerable link in network information resources. Phreakers typically perpetuate toll fraud by hacking into PBX systems as well as common carrier facilities to make illegal long distance telephone calls.

Piggybacking – Can assume a number of forms in modern computer networks. In the computer environment, piggybacking allows unauthorized users to gain access to computing time and applications by using or by mimicking the identity of an authorized user. Voice mail systems are another popular venue for piggybacking. In this instance, unused voice mail boxes are secured by hackers and used to distribute authorization codes, credit card numbers, and other illegally obtained data.

Point-to-point connection (PPP) – Supports multiple user access to the Internet. PPP can also provide connectivity to both asynchronous and synchronous devices. PPP also provides a high degree of flow control and error detection.

Post Office Protocol 3 (POP3) – Allows users to read e-mail in their mailboxes or server. Using POP3, a user's e-mail is transferred to the user client and is deleted from the server.

Primary Rate Interface (PRI) – The ISDN PRI supports data transfer rates starting at either 1.544 Mbps (twenty-three B and one D channels) in the United States and 2.048 Mbps (thirty B and one D channels) in other international ISDN settings.

Private network – Typically maintained by corporate, academic, or other privately held entities. These networks may be circuit switched, packet switched, client/server, peer-to-peer, and other current networking configurations.

Problem management – Concerned with the detection, prevention, and prediction of network outages, faults, or other disruptive events.

Protocol – A set of rules that governs the conditions for data transmission. Data rate, frame size, and other elements are often specified as part of a given protocol.

Public switched telephone network (PSTN) – Made up of telecommunication facilities provided by common carriers. The combination of local and long distance service (LDS) collectively make up the infrastructure of our public network.

Pulse code modulation (PCM) – PCM and TDM are used to transport analog data signals over digital facilities. The input analog signal (4 KHz) is sampled at regular intervals at a rate that is twice that of the highest frequency (8,000 times per second with each sample unit being 8 bits in length). The PCM process produces a data rate of 64,000 bits per second. Using TDM a T1 line with a rated data speed of 1.54 Mbps can be divided into 24 voice grade or 23 data channels capable of transmitting 64,000 bps. The PCM process involves four steps:

1. Analog data is input to a codec or digitizer.
2. The output of step one is a pulse amplitude modulation (PAM). T1 multiplexers are used to combine both analog and digital input signals for transmission over digital facilities. One of the major benefits of digital transmission includes its inherent ability to provide a common platform for the integration of various types of data.
3. The PAM sample is quantized or assigned a binary code.
4. The output of step three is an encoded PCM data stream.

R

Remote diagnosis – Network access to large medical databases and real-time access to medical experts around the world make it possible to provide consultation and diagnosis for a patient based on transmitted information.

Ring – In this topology, stations or nodes are attached to the network in a circular fashion.

Rotational delay – The times it takes for the read/write heads to be positioned over the desired data and instructions.

S

Secondary storage – Provides a means for storing programs and user data. In addition to providing reliable, high-capacity storage for data and program instructions, various forms of secondary storage are used extensively for backup and archival purposes.

Sectors – Tracks on a hard drive are divided into fixed length segments made up of blocks that contain data and instructions.

Secured electronic transactions (SET) – Based on SSL protocols for providing security in the Internet environment. Through the collaborative activities of companies such as MasterCard and Visa, SET has emerged as the de facto standard for providing security for Internet-based transactions. SET uses digital certificates that are issued by banks and other financial institutions. Remember, the certificate is the file that contains information used to authenticate the identity of the user.

Secure sockets layer (SSL) – Using RSA public key encryption, inter-layer communication between protocols and client and servers is protected. Data and other control information is exchanged between the aforementioned entities in what is known as *sockets*. Essentially, data integrity and security is built into the program layer of the application and the Internet. For example, Netscape uses SSL as a means of managing the security and confidentiality of data transmitted and received by this popular Web browser. Users are prompted when they are about to send to, or receive data from, an unsecured site.

Security management – Protection of data and the network from unauthorized access, manipulation and destruction, and other potentially harmful and often illegal activities.

Seek time – The time it takes to find the appropriate track on the hard drive.

Server – Many definitions have been associated with the term *server*. Within this context, server refers to a computer program that provides services on request to attached or connected clients.

Simple mail transport protocol (SMTP) – Part of the TCP/IP protocol suite that manages the transmission and reception of e-mail messages between users. SMTP ensures that the appropriate messages are routed to the appropriate users.

Simplex – In data transmission, a single path is used to either transmit or receive data. This unidirectional flow of data across a link is called simplex.

Single line Internet protocol (SLIP) – Another means of gaining dial access to the Internet. SLIP connections are typically slower than those provided by PPP and are essentially no-frills connections to the Internet.

Small Computer System Interface (SCSI) – Offers flexibility in microcomputer input/output options. By using a SCSI card, up to seven devices can be interconnected using a single computer expansion slot. SCSI devices such as Iomega's JAZ and ZIP drives, hard drives, digital audio tape (DAT), and CD-ROM can be added to existing computer configurations with ease. There are several standards specified for SCSI. SCSI-1 is first generation 8-bit SCSI technology that supported transfer rates of up to 4 MB. SCSI-2 provides 10 MB transfer rate using 16-bit architecture. More recently, wide SCSI or ultrawide SCSI-2 provides both 16- and 32-bit support with data transfer rates of 20 MB and 40 MB per second respectively.

Smileys – Used to temper the tone of an e-mail message. A few smileys are included for your review:

:-) smile

:->sarcasm

:-) laughing tears

;-) wink

:-<sad

Star – In this topology, stations and other devices are attached to the network where the file server may serve as the central point and all nodes and devices are attached to it independently.

Switched 56 – Was one of the earliest forms of switched digital transmission. Switched 56 remains a viable alternative for providing connectivity between customer facilities and production facilities. Also known as DS0.

Switching – In this context switching refers to routing decisions that are made by intelligent nodes within the network. The PSTN uses circuit switching. Networks such as Dow Jones uses packet switching. E-mail is a form of message switching.

Symmetric Digital Subscriber Line (SDSL) – Of all of the XDSL offerings, SDSL is perhaps the most versatile. SDSL provides an excellent platform for integrating variable rate digital transmission. Plain Old Telephone Service (POTS) and symmetric bidirectional high-speed traffic can be transported on the same line. SDSL provides variable high-speed transport ranging from 160 Kb to 2048 Mbps.

Synchronous – A mechanism used to manage the rate and flow of data as it is carried across a link. Clocks in communicating entities are synchronized in order to provide clocking or flow control.

Synchronous Optical Network (SONET) – A fiber-optic-based digital transmission offering providing transmission bundles ranging from 51.86 Mbps to 9953.28 Mbps. SONET represents an enhancement over other forms of high-speed digital transmission. SONET is based on an international standard called Synchronous Digital Hierarchy (SDH), providing users with a standardized digital interface and intelligent networking support.

Synchronous transmission – Instead of using start and stop bits to establish and maintain timing, clocks in modems and other networking devices are used. By grouping characters into blocks, less control information is required to send more data across the line.

T

Telemedicine – A number of processes have been identified under the heading of telemedicine. They include the transfer and manipulation of radiological and histological data, remote consultation, remote diagnosis, remote decision support, and diagnostic expertise.

Telepresence – Described as a system and associated applications that allow medical practitioners to share expertise, problem-solving skills, and sensory motor facilities with others at remote locations. Essentially, a doctor is able to perform some procedure or task at a remote location while having the sense of actually being there. This is achieved by providing sensory motor cues and feedback to the physician. The doctor is said to be virtually present in the remote location.

Teleradiology – The ability to transmit x-ray, MRI, and other visual information in digital format has been facilitated by the use of T1, ISDN, and other high-speed digital transmission facilities. The original or hard copy versions of x-rays and other medical images can be digitized using specialized scanning technology. Once digitized this information can be transmitted to remote sites instantaneously.

Telnet – This protocol allows users to log onto remote hosts or computers transparently. The user's machine becomes or appears as if it were a locally attached terminal.

Time division multiple access (TDMA) – The total available wireless bandwidth is divided into time slots. Each user is assigned a discrete time slot, permitting multiple users to share the same bandwidth sequentially.

Time division multiplexing (TDM) – By assigning the packets or frames from each user device a time slot, a single line can carry multiple user traffic. The use of time slots to carry multiple signals is called TDM.

Token passing – Another way to provide access to network resources would be to use a specialized series of bits called a *token*. Devices wishing to transmit must first seize the token before beginning information transfer.

T1 – A broadband digital transmission service provided by various common carriers. T1 service was first offered as a commercial service by AT&T in 1983. Using digital transmission facilities and TDM, T carrier provides digital data rates from 1.544 Mbps to 274.176 Mbps.

Tracks – Concentric circles on a cylinder of a hard drive.

Transaction validation – Concerned with ensuring that all Internet transactions occurred between valid trading partners, using electronic or other currency, and that the consumer receives what was purchased and the supplier receives remuneration for items purchased.

Tree – Stations are attached to the network in a perpendicular fashion. In short, several buses are arranged in a hierarchical fashion.

Trojan horse – A computer virus or routine that associates itself with a valid application program or software. The Trojan horse is far from harmless. Oftentimes, computer passwords and other protected information are located and stolen. The Trojan horse is also used to alter application programs, making them easier to copy or access.

U-V

Unguided or wireless facilities – Wireless communications implemented in the form of various radio frequency-based technologies have long served as an adjunct to wired networks. Wireless communications offers flexibility implementation over areas where it would be impossible to lay cable in highly developed metropolitan areas or rough terrain. While there are many different types of wireless communications in existence, we will examine microwave, cellular, satellite, and personal communications systems.

Very High Bit Rate Digital Subscriber Line (VDSL) – Provides a range of transmission offerings based on the distance between the subscriber and the central office. Essentially, the closer the subscriber is to the central office the faster the data transfer rate.

Voice mail – Another form of asynchronous messaging. Voice mail can be implemented using a card in the computer, a standalone voice mail system, or purchased as a service from a common carrier. Voice mail is an extension or type of automated attendant with a more personal touch. Usually a caller is greeted by the voice of the person whose mailbox was reached.

W-X

Wired facilities – Physical links between attached nodes and devices. Copper twisted pair, coaxial cable, and fiber optics are the most commonly used wired facilities.

Worm – A type of computer virus or program that infects network resources by replicating itself. Worms typically propagate on hard drives, secondary storage, and memory, having the cumulative impact of eventually crashing the system.

XDSL – The term used to describe a class of digital subscriber lines. The need to rapidly deploy digital transmission while minimizing costs associated with network upgrades has breathed new life into existing copper twisted pair used as part of the local portion of the PSTN.

Index

A

B

C

D

E

F

G

H

I

J

L

M

N

O

P

V

W